COMPUTERIZED

MANAGEMENT

OF MULTIPLE

SMALL PROJECTS

COST ENGINEERING

A Series of Reference Books and Textbooks

Editor

KENNETH K. HUMPHREYS

American Association of Cost Engineers
Morgantown, West Virginia

Additional Volumes in Preparation

COMPUTERIZED

MANAGEMENT

OF MULTIPLE

SMALL PROJECTS

RICHARD E. WESTNEY
Spectrum Consultants International, Inc.
Houston, Texas

Marcel Dekker, Inc.   New York • Basel • Hong Kong

Library of Congress Cataloging-in-Publication Data

Westney, Richard E.
    Computerized management of multiple small projects : planning, task &
    resource scheduling, estimating, design optimization, and project control /
    Richard E. Westney.
        p.   cm. -- (Cost engineering ; 17)
    Includes bibliographical references and index.
    ISBN 0-8247-8645-9 (acid-free paper)
    1. Engineering--Management--Data processing. 2. Industrial
    project management--Data processing.   I. Title.   II. Series: Cost
    engineering (Marcel Dekker, Inc.) ; 17.
    TA190.W479   1992
    658.4'04'0285--dc20   92-165
                                                                              CIP

This book is printed on acid-free paper.

MARCEL DEKKER, INC.
270 Madison Avenue, New York, New York 10016

Current printing (last digit):
10 9 8 7 6 5 4 3 2

PRINTED IN THE UNITED STATES OF AMERICA

For Marilynn, Laura, and Jill

Preface

The words "project management" conjure up images of hardhats, bull-dozers, and determined faces studying a construction drawing spread out on the hood of a dusty pickup. Many people still feel that, unless they are managing large projects, the techniques of project management have little relevance.

Until recently, that may have been true. Now everything has changed. Personal computers (PCs) have made project management methods relevant and extremely useful to anyone whose job performance is measured, at least in part, by the ability to complete projects on time and on budget.

Today, any job that has a deadline with a set of objectives to be met is a "project." Anyone who is responsible for getting projects done is a "project manager." Of course, that includes just about everyone. Who among us does not have deadlines to meet, budgets to estimate and stay within, and resources, including people and equipment, to schedule and coordinate? How many of us have several projects to handle at once? How many have to handle a situation where many projects compete for the attention of a fixed pool of resources? How many of us try to manage projects in a "downsized" organization with too few people, too many projects, and rapidly changing priorities? If any of this sounds familiar, then you are the person for whom this book was written.

Of course, we can easily conclude that just about everyone has project management as part of his or her job. This part requires planning, decision making, resource scheduling, estimating, performance measurement, and control. It is a difficult part too, especially when multiple projects are competing for the same pool of resources. And just about everyone can be better at project management, through the effective use of up-to-date tools.

So project management tools, correctly adapted to multiple small projects, belong in the arsenal of every professional person. The purpose of this book is to explain what these tools are, how they work, and how to use them.

Among the tools are the rich selection of sophisticated project-management software programs for PCs. In spite of their very low cost, today's project management programs provide features and capabilities that even mainframe users could only dream about a few years ago. Many programs have features and capabilities specifically for multiple small projects. But to gain the benefits that project management software can provide, the user must have a solid grasp of the relevant project management principles—the principles that are explained by this book.

The reader of this book could be a person who does project work, such as an engineer, scientist, researcher, foreman, accountant, or lawyer, etc. It could be someone who supervises the people and equipment involved in project work, such as a supervisor, resource manager, or coordinator. Chances are, that person currently finds project management to be more frustrating than satisfying. A project that has cost overruns or that finishes late seems to speak loudly about the competence of the person who managed it—even if the result was not necessarily that person's fault.

The type of small project that this book addresses can be simply defined as one that lacks a full-time project management team of specialists. In other words, management of the "small" project is usually a part-time activity by someone for whom project management is not a career specialty. The small project often must compete with other projects for the services of a fixed pool of resources. Because management perceives its cost as "small," the project may not get the attention it deserves although it may be far more important to the company than its cost would suggest. Examples of multiple small projects are maintenance, new product development, plant improvements, engineering work, research, and administrative efforts.

This book is based on an earlier text by the same author, *Managing the Engineering and Construction of Small Projects*. Readers of the earlier book will find that the focus has been shifted to multiple projects, and extra chapters added to deal with this and other timely subjects, such as value engineering and design optimization, assignment scheduling, and the use of personal computers.

Think of this book as a catalog of tools for small multiple-project management. After reading it, you will know what tools are available to you, and how to use them. You can be assured that every concept, method, and recommendation in this book is practical and proven. They are based on successful applications in actual multiple-project situations in many companies in diverse industries. These methods have been presented in public and in-house training courses, to many thousands of professionals from over 50 countries and from every imaginable industry since 1980.

These methods work—and they will work for you. Used correctly, these multiple-project management tools can make the project management part of your job a satisfying effort.

Richard E. Westney

Contents

I
DEFINING THE PROBLEM AND ELEMENTS OF THE SOLUTION

1

The Big Problem
of Small Projects

INTRODUCTION

For those of us who work with them, small projects mean big problems. Unlike our colleagues working on big projects, we have to cope with special problems like handling many projects at once, working in a production environment, and being on our own in the work we do. There are, fortunately, some proven project management concepts and methods which can be adapted to fit the small project environment. Before defining a solution to the small project problem, we must first define the problem in the most specific terms, and that is the purpose of this chapter. Although small projects exist in widely different circumstances, they all share certain common elements in both the problem and solution.

What Is a "Small Project"?

The word "project" can be used to describe an endeavor in which a number of tasks are performed in order to accomplish a particular objective. In the usual context of project management, projects are usually undertaken by an organization to achieve business objectives such as:

Maintenance of production capacity
Increased production capacity

Compliance with environmental requirements
Performance of research for new product development
Provision of engineering, mechanical, or construction services
Profit generation from the project work itself

We can see that the determination of what a "small" project is, is really determined by the environment in which the project takes place. However, we can define some guidelines as to what we mean by a small project. Small projects, in the context of this book, have one or more of the following characteristics:

Cost levels from $5,000 to $50,000,000
Cost levels less than 5% of annual budget for projects
Numerous other similar projects take place concurrently
Labor and equipment resources shared with other projects
The company doing the project is, itself, small
Project management efforts are part-time

Some examples of small projects are:

Plant maintenance
Research
Computer system development
Plant additions
Engineering
New product development
Plant modifications or improvements
Light construction
Projects to assure compliance with environmental requirements
Utility system outages
Administration

SMALL PROJECTS HAVE BIG PROBLEMS!

Size Belies Importance

Most people involved in project management (PM) would consider the types of projects listed above to be quite straightforward: to be not nearly so difficult as the multi-billion dollar "super-projects" to which so much attention has been paid.

The fact is, small projects can be just as important to the company involved as the larger projects and sometimes even more important. For example, a "turnaround" project, in which a critical manufacturing or process unit is shut down and overhauled in the absolute minimum time, can have a major impact on the plant's profitability if it takes too long and causes valuable production to be lost. Often the timing of the introduction of a new product is of critical importance, and depends on a number of engineering and construction projects to be completed on time.

So the value of successfully completing the small project can be far greater than the cost of the project itself, and the importance of the small project to the plant should not be underestimated just because the cost is small. If the company doing the project is itself small, the project may represent a major investment to that company.

Small projects are also important because of their increasing cost and complexity. As industries develop, and inflation continues, projects tend to become more complex and costly, making many types of projects suitable candidates for a more sophisticated approach to project management than had previously been taken.

The total cost of multiple small projects is often not small at all. In a large plant, such as an oil refinery or steel mill, the individual small project may represent an insigificant sum, but the aggregate cost of all the small projects done each year may be significant indeed. For example, the cost of maintenance each year often exceeds the expenditures for large capital projects and, unlike a large project which lasts a few years, maintenance work goes on continuously. So, if the total program of small projects is considered, it generally represents a project of significant size and complexity!

Perhaps the most difficult aspect of managing small projects is the problem of dealing with many projects at once. This certainly is a problem that the large projects do not have. In the small-project environment, many project engineers and maintenance managers must handle 20 or more projects at once, some of which are in the design and procurement stage, some of which are under construction, and some of which are just being started up. And it must be remembered that many project management activities must be performed regardless of the project's size. The basic problems of small projects are shown in Table 1.1

As illustrated by Table 1.1, in most organizations we find the paradoxical situation in which small projects, which have the toughest management problems, get the least attention. The reasons for this situation are described below.

Many Small Projects Exist in a Production Environment

Most small projects, involving maintenance, improvements, etc., take place in an operating plant of some sort. This facility operates for one

Table 1.1   Managing Small vs. Large Projects

	50 M$ Project Expenditure	
	Large	Small
No. projects	1	100
No. estimates	1	100
No. schedules	1	100
No. purchase orders and subcontracts	100-200	500-1000
Avg. project duration	3 yrs.	6 months
Full-time PM team	YES	NO
Formal PM control procedures	YES	Unlikely

reason: to make a profit by producing the maximum amount of on-spec product. Everything about the plant is dedicated to this one goal: its organization, procedures, priorities, and expertise.

Because the top priority is production, the small projects required to keep the plant running are often considered, at best, a "necessary evil". This typical situation finds the manager of multiple small projects constantly scrambling for the people, materials, equipment, management attention, cash, and even time required to get projects done. Many of the people on whom he or she must depend will perform project work only as a part-time, low-priority task. The production environment is also one in which things frequently change, as breakdowns and other unforeseen crises divert attention and manpower to unplanned but highly critical work.

Organization: Not Designed for Projects

As one might expect, the plant organization within which the small projects must be run is generally not designed for project management. The plant organization is, of course, intended to insure production, and must cope with the project work which is required in the best way it can.

Frequently, the manager of multiple small projects must communicate with, and draw support from, such organizational functions as:

Engineering, drafting
Purchasing, warehouse
Construction, maintenance
Project engineering, planning, estimating
Accounting
Upper management

The manager of multiple small projects often finds him- or herself in the classic dilemma of having responsibility without the authority to back it up. Since project work is apt to be a part-time, low-priority job for these various support groups, the problem of getting the work done can become acute. And, the problems are often compounded by the fact that he or she has had little opportunity, at school or work, to be trained in the principles and methods of project management.

Perhaps because of these organizational problems, small projects often suffer from a lack of the formal procedures, methods, and data which are available to larger projects. In spite of the ease with which actual cost and schedule data could be gathered, planning and estimating are often done without the benefit of a good database, and often without experienced planners and estimators. In addition, this problem is frequently made worse by the lack of time available to do the planning and estimating in the first place.

Small Projects: A Special Problem of Control

This lack of a sound plan and estimate naturally leads to problems in project control. Compounding these problems are the special aspects of the small project such as:

Short project life: This leaves little time to gather data, identify problems and correct them.

Shared responsibility: This makes it difficult to obtain commitments and enforce accountability among individuals and departments.

Problems in obtaining actual data: These problems lead to inadequate reporting resulting in a lack of the information required for effective control.

Many projects to be controlled simultaneously: The number of projects adds complexity as some projects will be in early stages while others are near completion. The sheer number of projects handled can often be a problem: Many facilities have hundreds of small projects in progress simultaneously.

Why Standard Approaches Don't Work

Many managers of multiple small projects have tried the procedures and techniques developed for large projects in an attempt to improve the management of the smaller projects. These attempts are often unsuccessful because large-project management techniques require a project team with specialists in, for example, planning and cost engineering, and computer applications.

Standard approaches are usually based upon a detailed plan and cost estimate that can be used as a basis for control. Since most small projects do not receive a high enough level of effort in the planning and estimating stage, there is often no basis for exercising project control in the usual way. Also, standard approaches are not suited for the short timeframe within which small projects are executed, the division of responsibility, or the problem of dealing with many projects at once.

The fact that most small projects are executed at plant-level is also a key reason why standard approaches don't work. Plant operations impose a great many constraints on a small project that do not exist for large, grass-roots projects. Access is likely to be restricted to the work area, hot-work permits will probably be required, construction and maintenance personnel must work among operations personnel whose work has priority, and the unpredictable nature of plant operations is likely to cause numerous changes to the scheduled access to the unit and the availability of personnel.

Perhaps the main reason why standard approaches don't work for small projects is that many small projects are "revamps". What is a revamp project? A revamp project is a change to an existing facility. Such changes are usually made to improve the unit's performance in some way. Examples include:

Debottlenecking projects to increase the unit's capacity by replacing piping or equipment that is currently limiting performance

Changes to improve safety, operability, or maintainability by adding such items as lighting, platforms and stairways, piping and valving, alloy materials, redundant or oversized equipment, etc.

Additional facilities to maintain or improve operations

Facilities to assure compliance with present or anticipated environmental requirements

Modernization projects

Major maintenance projects often share the special problems of revamps. These special problems include:

Complications caused by working in an operating plant

Congestion in the work area

Lack of access to the work area, which often creates the need to perform work in a non-optimum sequence

Interference from normal plant operations

Inadequate design definition causing quantities of piping, steel, and electrical materials to overrun because the actual placement and routings are not defined until the construction takes place

Engineering manhours often overrun because the scope of work is difficult to define

Field manhours often overrun because both the scope and difficulty of the work have not been properly defined

Costs often overrun, not only because of increased quantities and manhours, but also because overtime or shift work may be required to get the work done when access to the unit is permitted, or to get the increased scope of work completed on time

Each project tends to be thought of as unique, with little in common with previous projects and therefore with no easy way to define it

Turnaround projects (or "outages"), in which a unit is shut down for major maintenance work that is carried out on a "crash" basis, tend to be better planned and controlled than more routine revamp and maintenance work. Because they usually get attention and support from top management, and because there are no conflicts with an operating plant, turnaround projects are more like large projects for which standard approaches will work well.

Small Projects Need Special Management Techniques

It is certainly evident that small projects have special problems that require special project management techniques. Any project management technique, to be successful in the small project environment, must be able to:

Handle many projects at once

Be used effectively by project personnel with no training or experience in planning, cost estimating, or project control

Cope with the short timeframe of small projects

Simplify the organizational interfaces

Handle the complexities of work in an operating plant

Provide a basis for accumulating cost and schedule data for use in planning and estimating future projects

Improve the multiple project manager's capability for estimating, planning, scheduling, resource management, expenditure forecasting and

control, capture and analysis of project data, and preparation of management reports
Provide a consistent approach to be applied to each project and by each project manager
Be adaptable to existing procedures

If it were easy to provide such a management system, there would be no need for this book. It is, however, possible to adapt certain proven project management techniques to the special problems of small projects. That is what the remainder of this book is about.

EXAMPLES OF MULTIPLE SMALL PROJECTS

Because the environment in which small projects are executed varies greatly, the following projects are presented to highlight the special aspects of small projects in various typical environments.

Special Problems in Large Manufacturing Plants

Typical small projects include maintenance, turnarounds, plant additions and plant improvements. The problems of the manufacturing facility center primarily around the overwhelming importance of production and the resulting fluctuations in priorities, as well as the availability of materials, labor, engineering services, and access to the unit. Projects in manufacturing plants often experience large increases in the scope of work, as the difficulties involved may not be apparent until the work is progressed to a certain point. Also, major maintenance projects, such as turnarounds, operate under tremendous time pressures but the scope of work may be impossible to define until the unit is shut down and the reactor vessels or other equipment items are opened up. (Many maintenance crews have opened up a pressure vessel to find all its internals lying peacefully at the bottom.)

Special Problems in Large Corporations

There are many small projects that are carried out by large corporations that do not involve manufacturing, such as engineering and research work. Although they do not involve construction, these projects share the same problems of the nonproject environment and multiproject management. A frequent question in an engineering department in which 20 engineers are each working on five projects is apt to be, "To whom should this new work assignment be given?'' Engineering projects are also difficult in that physical progress is harder to measure than in construction work, and that the effect of changes can be far greater than most people realize.

Research projects are characterized by the very high quality of the technical, manufacturing and construction work, and the very low (usually zero) profit results. A research project constantly weighs that the trade-offs between the technical quality necessary to assure good results, and the budget limitations.

Special Problems in Small Companies

Small companies are directly or indirectly involved in most small projects. To a small company, of course, a project that is small to the client may be the biggest one the company has ever handled. In spite of this difference in perspective, the small-company project is usually done without the highly trained personnel or sophisticated systems that large companies can afford. Small-company projects can also be critical if they represent a lump-sum contract on which the profit margin is important. Perhaps no one is more attuned to cost control than the manager of a ''hard-money'' contract.

Special Problems of Projects in Remote Locations

Many small projects take place in remote locations such as the Arctic, the desert, the jungle, the mountains, or offshore. Of course, the most important and difficult aspects of these remote projects are logistical; getting men and materials to the site, communications, and difficult working conditions at the site. Resource planning and control is therefore particularly important on this type of project.

CHAPTER SUMMARY

Effective project management, like other management functions, is essentially the ability to do three things well: leading the people doing the work, making decisions, and, above all, communicating. For the small project environment, these things are both essential and difficult.

The special problems of small projects are significant and difficult to solve. They are, however, worth solving because of the often overlooked importance of the projects to the companies involved. Although specific situations vary, all small projects need proper planning, scheduling, contracting, and management, and the organizational problems can be addressed by techniques that provide a basis for communication, commitment, and control. Fortunately, the methods that are available to manage small projects also provide the necessary basis for resolving the organizational problems as well. These methods are derived from the basic concepts discussed in Chapter 2.

2
Basic Concepts of Small Project Management

INTRODUCTION

In Chapter 1 we discussed the special problems of small projects, explored some of the reasons why standard approaches to project management do not work in the small project environment, and outlined some of the performance criteria we apply to any management tools we propose. In this chapter we define the basic concepts that will then be developed into the specific project management tools to be described later in this book.

CRITERIA FOR SMALL PROJECT MANAGEMENT TECHNIQUES

From the problem definition provided in Chapter 1, we can see that, in order to handle the special problems associated with small projects, an effective project management technique should have:

1. *A standard approach.* Although each small project is apt to vary significantly from all others, all projects need to be handled with a standard approach. This requirement is to assure efficiency and consistency on the part of all those who are associated with the management of a project (including corporate management). A standard approach minimizes the time and effort required and is also much more likely to be used and

understood. However, a standard approach must be flexible enough to be effective under all circumstances that may be encountered on diverse projects.

   2.   *Simple systems and techniques.* Although the small project management problem is complex, the diverse nature of the personnel who will be using (or exposed to) the project-management techniques and the need for ease of training requires that the techniques involved be simple to use and understand.

   3.   *Fast response.* The short duration of the typical small project indicates a need for techniques that provide information and answers quickly and effectively. This requirement also stems from the rapidly changing environment in which many small projects exist: any change in priorities, dates, scope of work, or group workload can impact the schedule and resource availability of the small project. This indicates a requirement for flexibility, to be able to rapidly adjust protect plans as the situation changes.

   These requirements describe a systematic, consistent, and effective project management and control system that will perform well in the multiple small-project environment. This system can be constructed by using the basic concepts defined below.

BASIC CONCEPTS OF SMALL PROJECT MANAGEMENT

There are four basic concepts that form the basis of an effective small project management technique:

Integration of cost, time, and resources
Use of formal planning methods
Use of project models
Effective use of personal computers

   These basic concepts, which are well suited to any project, are particularly appropriate to the small project environment as they result in significant improvements in management efficiency and effectiveness. This can be seen as each of these basic concepts is discussed in detail below as follows.

Integration of Cost, Time, and Resources

An integrated approach has been demonstrated to be the only truly valid and effective technique for project evaluation and control. By an integrated approach we mean the use of techniques that recognize the dependency of project management variables. This integrated approach is used by all state-of-the-art project management software. Integration of project-management parameters is a very powerful concept, worth careful study and consideration. Although developed and proven in large-project applications, the enhanced efficiency and effectiveness that results from the integrated approach has great relevance to small-project applications. Let us begin our study of integrated-project control by examining the reasons for its greatly increased effectiveness.

THE VARIABLES OF PROJECT MANAGEMENT

Project management is the process of definition and manipulation of "project variables" such that the objectives of the project are achieved in an optimum way. In general, it can be said that there are four categories of project variables:

The *cost* of the project
The *time* it takes to perform the tasks that comprise the project
The *resources* required to complete the tasks
The standards of *quality* that are applied

Even on a small project, these variables are explicitly or implicitly controlled, and the results of the project—how much it cost, how quickly it was completed, how efficiently the company's resources were utilized, and how well the resulting facilities operated—all depend on how well the task of controlling these variables was performed. To put it another way, we can say that we are seeking the "optimum blend" of these variables.

Many managers of small projects overlook the actual objective of a project from corporate management's point of view. From that lofty perspective, it can be easily seen that the "bottom line" as to the success or failure of a given project or group of projects is, in fact, the "bottom line". That is, every project, large or small, exists for one reason alone, and that is to maintain or improve the company's profits. The contribution of a project to company profitability is, of course, determined by the relationship of the variables listed above. Let us look at each one of them in more detail:

Cost

The cost of a project can be described from several points of view:

The *investment cost*, which is the total investment in the project
Expenditures, which are the cash required each month during the life of
    the project

Operating costs, which are required to maintain and operate the facility
over its useful life

As managers of multiple small projects, we are concerned with controlling the investment cost as well as advising the accounting department
on how much cash they can expect to be called on to pay out. Our objective in cost control is, however, not necessarily to minimize the investment
cost. It is to insure that the cost implications of all the many decisions
and developments that occur are properly considered so that the optimum
cost may be achieved. For example, a design change may be proposed in
order to improve maintainability and thereby reduce operating cost. The
question is: "Is it worth the increased investment cost?" That question
can, of course, only be answered if the effect on investment cost is known,
and cost control is the means by which we assure that the increased cost
is defined in a timely manner such that it becomes part of the decision as
to whether or not to proceed with the change. So the optimum investment
cost may not necessarily be the minimum cost.

In another typical situation, we might be faced with an unexpected
delay in the receipt of materials. Our choices are to let the progress slip,
pay a premium to expedite delivery, or accept a later delivery but work
some overtime to maintain the scheduled completion date. Clearly, all
of these alternatives have some effect on the cost of the project. Once
again, our cost-control efforts are directed toward assuring, not that the
minimum cost approach is taken, but rather that the impact on investment
cost is identified and is incorporated into the decision process.

In both of the above cases we might ask, "Why is the lowest cost alternative not necessarily the best one?" The answer is simply that the best
alternative is the one that has the most favorable impact on profitability,
and profitability is affected by all four of our project variables; cost, time,
resources, and quality.

Time

Time, like money, materials, or manpower, is a resource that we manage
to achieve our project's objectives. Projects vary enormously in the importance that is placed on timing. In some cases, significant improvements
in profitability can be attained simply by accelerating the startup date by
one day. Or, in the case of projects that are required by legislation, such
as environmental improvements, the company's failure to achieve startup
by the required data can result in legal penalties that must be avoided by
any cost.

We plan and control our utilization of time with the project schedule, in which we identify the points in time where various project activities will start and finish. Many project managers view schedules with an attitude of "the sooner the better", and frequently make great efforts to accelerate the start and/or completion of activities based on the assumption that some good must always come out of it. However, the cost implications of such efforts are often overlooked.

The Relationship Between Time and Cost

"Time is money." We all accept that old truism, and it is definitely true for projects. Figure 2.1 shows a curve describing the relationship between time and cost. If we start at the point of minimum cost (i.e., maximum efficiency), and assume that it is desirable to accelerate the schedule we can define a number of ways to do that. Note that, in this discussion of time vs. cost, we are assuming that resources, such as manpower, are held constant.

One possible way to advance the schedule without adding manpower would be to have everyone work overtime. This results in higher costs-per-hour due to the necessity of paying a premium for the overtime hours, as well as the reduced productivity for work done during overtime. This

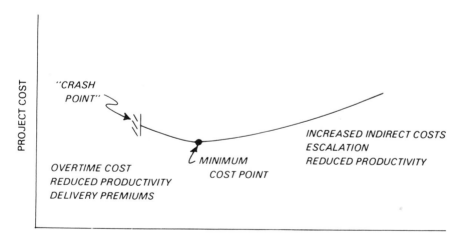

Figure 2.1   Relationship between time and cost. Manpower levels are held constant.

results in more work-hours of effort being expended. In return for these higher costs, some reduction in the time required to do the work can be expected.

Another way in which schedule acceleration can be bought is with the use of incentive payments to suppliers, fabricators, and field subcontractors who are able to beat their schedules. Sometimes these incentives take the form of awarding a contract based on the promised delivery or completion date, rather than the lowest price.

Thus we can see that, in general, a tighter schedule will result in increased costs due to various forms of premium payments and reduced productivity. There is, of course, an absolute minimum time in which the required work can be accomplished. This is often referred to as the "crash point." It would seem only fair that if schedule improvements tend to increase costs, then schedule slippages should result in cost savings. However, the various laws of nature, as well as those of Mr. Murphy, have conspired to assure that this is not the case. What happens to the project's cost when the schedule is extended?

Project costs are often described as having "direct" and "indirect" components. *Direct costs* are those costs that relate to activities which contribute directly to measurable progress. *Indirect costs,* then, are those costs that relate to other activities such as supervision, office services, warehousing, provision of construction equipment, etc. Many indirect costs are determined primarily by time, that is, they are incurred every day or month regardless of the amount of work that is done. Office accommodations and staff costs are good examples: the rent must be paid whether or not the office is busy. It can therefore be seen that, as a project extends past the point of maximum efficiency, the indirect costs will increase.

Productivity can also be affected when the schedule is extended, depending on the reason for the delay. If delays are the result of strikes, work stoppages, or other forms of labor unrest, there will generally be a loss in productivity resulting in more work-hours of effort being required to perform the work and an associated increase in cost. Other delays can be due to drawings or materials being delivered late, resulting in work being stopped and started, duplicated, or done out-of-sequence, all of which can result in lost productivity and increased cost. Even the psychological impact of this loss of "momentum" can have a pronounced effect on productivity.

We can conclude, therefore, that cost and schedule are inherently related. Actions taken to control schedule will have some effect on cost.

Actions taken to control cost will have some effect on schedule. An integrated approach deals with each of these project variables as separate terms in the same equation, recognizing that each affects the other and that both affect the project's success.

Resources

Project resources provide the means of accomplishing the work required to achieve the project's objectives, and it is the way that resources are managed which determines the project's eventual time and cost. In spite of its importance, this aspect of project management is frequently neglected. To those involved with small projects, resource management may seem to be a lot of extra work to do something that could be done intuitively. However, it is on the small project that resource management is, by far, the most critical since small projects have very little alternative work that can be done when a lack of labor or materials stops progress. Now, we shall look at how resources are related to cost and schedule.

The resources required to accomplish a project generally fall into the following categories:

Professional manpower (e.g., engineers)
Materials (e.g., pumps, piping, steel)
Field labor (e.g., welders)
Subcontracts (e.g., insulation)
Equipment (e.g., construction equipment)

Our task in project management is to see to it that the required quantity of each resource is available when needed, that it is used properly, and that it is demobilized promptly.

Why is resource management especially critical for small projects? Consider the degree of flexibility that exists on a large project. If an important material item is not delivered on time, if a key piece of construction equipment is suddenly unavailable, or if required crafts are not present in the desired quantity, there is always enough work to be done so that progress can be made in spite of these difficulties. Work in other areas (work with other equipment and crafts) can usually be progressed. But consider the plight of the small project in similar circumstances. There may very well be no other areas to work in, no other material to work on, no way to make progress without the necessary crafts and equipment. And, in addition, no time to waste since the time it takes to work out these resource problems may represent a major portion of the schedule.

Most small projects also exist in a multiple project environment in which each project must "compete" for attention and services from a fixed pool of resources. So resource management is indeed vital to the small project. To be successful, it must be done efficiently and in a way that recognizes the effect of resources upon project time and cost (see Chapter 4 for details).

The Relationship Between Cost, Time, and Resources

Let us now consider our cost vs. time discussion illustrated in Figure 2.1. What happens if we allow the resources applied to the project to vary as well? Figure 2.2 provides an illustration of the exaggerated effects brought about by resource variations.

Suppose that once again we wish to accelerate the schedule, and now we intend to apply the maximum amount of manpower possible. This will require three shifts working 24 hours per day, seven days a week, which will, in turn, add to costs for shift premiums. Further cost increases will result from reduced productivity due to coordination problems between shifts, lower productivity on evening and night shifts, and increased manpower density. However, these additional extra costs will result in our crash point being moved back even further as we achieve even greater schedule improvement. Should our project be delayed, we can reduce the cost penalties somewhat by reducing resource levels to the minimum that can be used effectively.

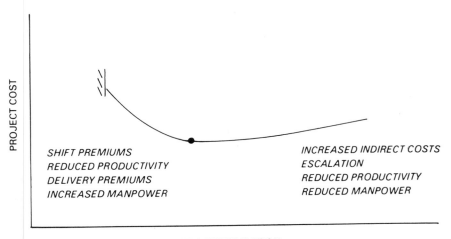

SHIFT PREMIUMS   INCREASED INDIRECT COSTS
REDUCED PRODUCTIVITY   ESCALATION
DELIVERY PREMIUMS   REDUCED PRODUCTIVITY
INCREASED MANPOWER   REDUCED MANPOWER

PROJECT DURATION

Figure 2.2   Relationship between time, cost, and resources. Manpower levels are varied.

From the above discussion it can be seen that cost, time, and resources must be managed in an integrated way if truly effective control is to be achieved. This approach is particularly appropriate to the small project that can so easily be delayed or overrun when manpower and materials are not available when required. The useful cost-time-resources curve need not be just theoretical: it can be drawn for any project using the technique of "project modeling", discussed later in this chapter. The integrated approach can produce substantial improvements in individual and organizational efficiency, making the project engineer's job easier and his or her efforts more effective.

Quality/Scope of Work

The level of quality of the work, often determines the scope of the design specified and the standards to which it will be built, and has a clear effect on cost and schedule. An increased level of quality usually results in increases in work, project cost and schedule length. And, as shown above, the optimum technical quality is that which results in the maximum return on investment, not necessarily the level that costs the most.

Although the importance of quality management is evident, it is a much misunderstood subject and is discussed further in Chapter 3. For example, quality control and quality assurance efforts are usually aimed at assuring that a certain level of quality is achieved, not that the design quality be optimized for cost, schedule, and return on investment. Also, quality control should start when the design work begins. The basic design for a project can be thought of as setting a minimum cost level. That is, given the design, there is a certain minimum project cost that must be incurred. Though changes are infinitely easier to make at the early stages of design, how often is there comprehensive design optimization on a project? In most cases, the design is completed, an estimate is prepared, the project is built, and the facility operates for years in which may or may not be the most profitable way.

The relationship between quality, cost, and schedule can be seen in Figure 2.3. As quality is increased, the cost and the time required are likely to increase as well.

Efficiency Gains from an Integrated Approach

In Chapter 1, the organizational problems associated with small projects were seen to include a lack of manpower to manage the project most efficiently. The integrated approach is particularly well-suited to multiple small projects because it avoids the duplication of effort that is often found in the standard approach to project management.

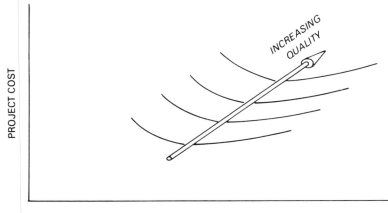

Figure 2.3 Relationship between time, cost, and quality.

Let us consider the two major phases through which any project, large or small, must pass. The first phase might be referred to as the "project evaluation" phase. In this phase, the basic issues which must be addressed include:

What will the project cost?
How long will it take to do it?
How will the work be done?
What resources are required to do it?
How much additional profit can be made?

Once the project is approved, it enters the "project execution" phase in which the design, procurement, and construction take place. During this phase, efforts are dedicated to the management and control of the project's cost, schedule, quality, and resource utilization. As we now know, our true objective is the optimum blend of these project parameters. Both phases consist of efforts by diverse groups involved in these functions:

Cost estimating
Planning and scheduling
Purchasing and material control
Quality assurance
Construction planning and supervision

Economic analysis
Accounting
Management

In most cases, these functions are performed separately, with meetings, memos, and phone calls being the primary modes of communication. Each function is apt to be performed using informally acquired data and unofficial methods. This practice can often result in inaccuracy, inconsistency, and inefficiency since there is a considerable duplication of effort, as well as excess time spent in communication. All the above functions are similar in that they involve tracking, analysis, forecasting, and reporting of project data, identification of problem areas for attention, and followup of any corrective action that may have been taken.

An integrated approach, then, provides us with enhanced efficiency as seen in Figure 2.4. Instead of each organizational function using its own data, the integrated approach allows those involved in the projects to use a central database and model of the project.

Use of Formal Planning Methods

The second of our basic concepts is the use of formal planning methods. Although virtually all large projects these days use planning networks,

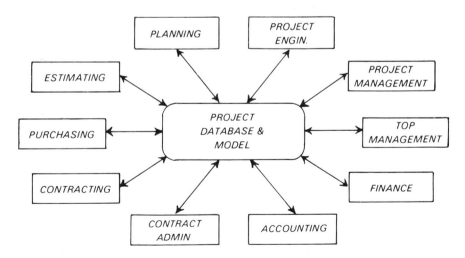

Figure 2.4   Distributed processing using integrated project data.

they are sometimes regarded as too cumbersome and complex for small-project use. There are, in fact, a number of compelling reasons for using networks on small projects (see also Chapter 3).

A Network Plan Is the Best Way to Represent a Project

Network plans have been used for the past several decades, and have gained much broader acceptance in the past 10 years or so, as computers became more available for project use. The growing use of network plans is due primarily to the fact that it is the only valid way to represent a project. The essence of a project is simply many different people doing many different things over a certain period of time, working toward a common goal. A network plan allows us to describe those people and their activities in a rational, quantitative, and mathematical way. In addition, it is a tool that can be used for many different applications such as scheduling, estimating, resource analysis, cashflow forecasting, and project control. To be effective for small projects, networks must be kept simple. If they are, they become an extremely useful tool.

Network Plans Must Be Simple

The human brain is probably incapable of understanding, and using effectively, a network plan of over 200 activities. In fact, as a general rule, 50 activities is probably a good maximum for a small project. Surprisingly, any project, no matter how large, can be represented by a simple network plan: it all depends on how much detail is included in each activity. We can, in fact, construct a hierarchy of network plans, as shown in Figure 2.5, in which a complete network at one level is represented by one activity at the next higher level. It also can be seen that the need for detailed information is, in general, inversely proportional to one's position in the organization: that is, the higher one's position, the less detail is required. Thanks to our ability to select a level of detail for our plan that matches the level of detail needed by the user, plans need never be too complex for effective use (see Chapter 3 for the use of network plans).

Network Plans Provide the Basis for Integration

There is, in addition, an important reason for using network plans that goes beyond the general benefits described above: *the plan provides the basis for integration of the cost, time, and resources required to complete the project.* In other words, the plan is our framework for integrating and manipulating these key project variables. The following is a description of how this process works (see Figure 2.6). For each activity on the plan we identify:

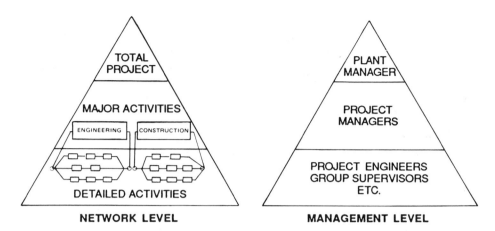

Figure 2.5 Hierarchy of network detail.

SCOPE: *PREPARE PIPING DESIGN DRAWINGS*

DURATION: *15 DAYS*

RESOURCES: *(1) PIPING DESIGNER*
             (3) DRAFTSMEN

COST: *(1 DESIGNER) (15 DAYS) (320 $/DAY)*   = *$ 4,800*
             (3 DRAFTSMEN) (15 DAYS) (250 $/DAY) = *$11,250*
                                                      $16,050

Figure 2.6 A resource and cost loaded network activity.

The *scope of work* defined by that activity (i.e., activity description)

The *time* available to perform that work (i.e., activity duration, expressed in working days)

The *resources required* (e.g., materials and labor) to perform that scope of work in the time available

The *cost* to provide those resources over that period of time

Having defined the cost, time, and resources for each activity such that they are consistent and in balance, we can say we have a resource- and cost-loaded network plan for the project. But why do all this for multiple small projects? The reason is that the fully integrated project plan becomes the basis for communication, consistency, coordination, commitment, and control. These five aspects of project management tend to be inadequate on small projects.

Communication: This is enhanced when general concepts, plans, and intentions are put in specific form and clearly documented.

Consistency: This is improved when each plan, estimate, and progress report is prepared using the same approach, regardless of who does it.

Coordination: This is improved by the ability to see what work needs to be done, as well as the consequences of the failure to perform that work. This is especially important in the functional- and matrix-style organizations in which most multiple small projects are done.

Commitment: This is obtained when an understanding is reached according to a specific definition of what is needed, when it is needed (and for how long), what resources are required and a specific agreement is made to provide it.

Control: This is possible when, as the project proceeds, an agreed-upon plan is available for comparison with what actually happens.

A final advantage of network plans is that they provide an excellent means of computerizing project information. Given the amount of data involved in multiple small projects, the need for quick results, and the lack of manpower available for data analysis and report preparation, it is evident that effective computer assistance to the multiple small project manager would be most welcome.

If we assume we have a fully resource- and cost-loaded network, and that we have set it up such that we can manipulate the data, we can say that we have created a "model" of the project.

Modeling

A project model is a complete mathematical and descriptive representation of the project, expressed in terms of the project variables of cost, time, resources, and scope of work. It can be input into a personal computer such that manipulation of the data will show the outcomes of various scenarios.

Why build such a model? To answer that question best, one should consider the better-known models used in industry. For example, we are all familiar with the econometric model used to predict inflation trends, gross national product, and other measures of economic performance, and most engineers have worked with computer models that run calculations to simulate what will happen under certain conditions. Models need not be computerized: consider the wind-tunnel model of an aircraft, used to simulate performance with different configurations, or the model basins used to test ship designs. What these models all have in common is the ability to simulate a real-life situation in order to see what will happen under a given set of conditions. The model is helpful because it would be impractical or impossible to do such a test in real life. For example, government economists don't experiment with the national economy (although there are times when it seems that they are), but they can test the outcome of various decisions by using their economic models. Similarly, it could be quite dangerous to use real aircraft to do configuration testing that can be done safely with wooden models in the wind tunnel.

But why model multiple projects, especially small ones? Because a multiple project plan is an exercise in the optimization of variables. These variables involve people in real-life situations in which they respond in a very complex way. In addition, a project involves the manipulation of a lot of data with results required quickly. The time-cost-resource curve shown in Figures 2.2 and 2.3 cannot be drawn other than by manipulating the variables with a project model. In fact, a project model is a fancy way of describing something very simple: a way of analyzing the project parameters of cost, time, resources and scope so we can record what is planned, what has happened, and what will happen. The model is how we accomplish integration.

Effective Use of Personal Computers

The last of the four basic concepts for small-project management is effective computerization. Much has been said, written, and done about computer applications, and, since personal computers have much to offer the multiple small project manager, this book will be no exception (see Chapter 15). Our task at this point is to examine the question of why computers are a practical tool for managing multiple small projects.

Benefits of Computer-Assisted Methods

No doubt most readers will have had extensive experience with computers and seen that they can, in almost any situation:

Handle a lot of data very quickly
Provide a standard approach and assure consistency
Be made to be quite simple for the user
Assimilate and organize data from diverse sources
Provide reliable storage of data for use by others.

In addition, we can see that an integrated, modeling approach is greatly enhanced by computer assistance.

From our definition of the problems of small projects, in Chapter 1, it is evident that many small-project problems could be alleviated by an effective computer system. Yet many companies today are still using manual techniques even when they are known to be inadequate. One of the many reasons for this is the fact that, in past years, computerization often turned out to add new problems rather than solve old ones. Yet there is no doubt that computer systems have something to offer our hectic, disorganized world of small projects. The questions are: what can be accomplished, and how can it be done? The answers begin to be found with recognition of certain trends in computer technology which have given us some new developments that have profound implications for project management.

Developments in Computer Technology for Project Management

During the past five years there have been developments in both hardware and software that have drastically changed our ability to use computers effectively in the project environment. The most notable advances are:

1. *Distributed processing.* The ability to bring computer power right to the person using it. This provides instant response, with direct input by the user and immediate output from the system, and it means that a computer-based solution to a problem can be implemented quickly and easily. Distributed processing allows all those who might be associated with a project to access and update information in the central project model and database shown in Figure 2.4.

2. *Interactive operation.* The ability to input information or commands via a keyboard and receive an immediate response to the instructions given. In the multiple small-project environment, this means that the time required to capture, analyze, and present data is reduced to the point where one can control projects to an extent far beyond what was accomplished before.

3. *Lower cost.* With the advent of minicomputers, workstations, microcomputers, and personal computers, the cost of computer-based solutions has dropped to where it is almost insigificant. And, because

smaller systems are available off-the-shelf, computer-based solutions can be implemented, if necessary, in a matter of days. This is very helpful in the multiple small-project environment in which little money is available for project-management systems, and often, little time is available for implementation.

4. *Flexible software.* Consistent with the availability of interactive, command-driven systems, software has become available which allows the user to specify exactly what he or she wants. This is perfectly suited for multiple small-project management procedures and systems, which have to fit existing company practices that are unlikely to be changed to suit the project engineer's needs.

5. *Project management software.* During the past five years, the need for flexible, effective, and easy to use project-management systems has been recognized by vendors who now offer a large selection of software specifically designed for management of multiple small projects.

6. *User-friendly systems.* As more people have become system users, the systems themselves have become easier to use, i.e., more "friendly," easier to learn, and more tolerant of mistakes. Since the small-project environment generally has little time or budget for development and training, and since a variety of people need to have access to the system, user-friendliness is quite important.

CHAPTER SUMMARY

We have, in this chapter, defined the basic concepts of project management that will be developed throughout this book for application to the multiple small-project environment. Each of the remaining chapters explores an aspect of small-project management, and shows how these basic concepts are extended into practical tools for the manager of multiple small projects.

II

PLANNING, SCHEDULING AND ESTIMATING THE SINGLE SMALL PROJECT

3
Planning the Small Project

THE BENEFITS OF PLANNING

Why spend time thinking about the future when the day is filled just handling today's problems? Many project leaders feel they just don't have time to plan—and with good reason. Planning is often seen as difficult and time-consuming, with little tangible benefit. After all, we know that things never work out according to plan!

There are many good reasons for planning. Fortunately, today's project management software makes planning so easy and intuitive that little time is needed to plan. Let's examine some of the benefits of taking the time to plan a small project:

planning identifies problems: It predicts potential problems and provides
   sufficient time so that steps can be taken to prevent the problem from
   materializing
planning encourages communication: It provides a basis for communication between those who are responsible for the project results, and
   those who must provide the resources and services it requires
planning provides the basis for decision-making: It allows simulation of
   project data to examine the impact of variations on scope, resources,
   priorities, etc.

planning encourages proactive management: It provides a reason and a
basis for structured thinking before a project begins—resulting in a
well-controlled project in which problems are anticipated and events
are controlled

planning provides tools for project control: It forms the basis for all the
project management and control tools described in this book including: Critical Path Method (CPM), Program Evaluation & Review
Technique (PERT), resource scheduling, multiple-project resource and
task scheduling, cost estimating and cashflow forecasting, earned
value method for measuring performance, and cost and schedule
forecasting

planning provides the means to keep track of changes: It allows project
data to be easily updated to reflect actual or anticipated changes in
scope of work, design, resource availability and so on

*planning provides an unbiased basis for communication between client
and contractor*: It serves as a way for schedules and budgets to be
agreed upon and updated as the project develops

planning methods assure consistency among multiple projects: A classic
problem of multiple-project management is the lack of consistency
between methods used on various projects. Formal planning methods
provide a consistent approach, leading to greater overall efficiency
and effectiveness

a final benefit: Computer systems for project management pay off the
time and effort required to use them *by giving us useful information
we would not otherwise have.* Armed with this information, we can
manage strongly, proactively, and effectively—minimizing risk and
maximizing return.

Now let's explore the theory and practice of planning a small project in
the multiple-project environment. Bear in mind that, if multiple-project
management is to be effective, all projects, no matter how small, must
be planned and included in the multiple-project method.

WHAT IS PLANNING?

There are a great many expressions regarding planning that are often used
loosely, and it is therefore important to start with a clear definition of the
terminology that will be used.

"Planning" is the process of breaking a project down into specific
tasks, and defining the sequence in which those tasks can or must be performed. Planning is not the same as "scheduling." Scheduling is the process

of defining the dates on which each task will be performed, and thereby determining the start and completion dates for the project.

We could say that planning defines "what" is to be done, while scheduling defines "when" it is done. The resource schedule, which is discussed in Chapter 4, describes "how" the work is done. By clearly defining, before the work starts, just what we have to do, how we intend to do it, and when it will be done, we establish a basic agreement for the small project.

DEFINITION OF PLANNING TERMINOLOGY

"Critical Path Method" (CPM): CPM is a planning technique in which a plan is prepared and a sequence of activities is defined such that, if the completion of any one of them is delayed, the completion of the project will also be delayed. This sequence of critical activities is called the "critical path." The critical path is the longest sequence of activities and determines the minimum time in which the project can be completed.

"Program Evaluation and Review Technique" (PERT): Although CPM and PERT are often used synonomously, like planning and scheduling, they are, in fact, quite different. PERT is a technique that allows us to deal with the uncertainty of the duration or costs of the activities on our plan. Using PERT, we can define a most probable duration, an optimistic duration, and a pessimistic duration for each activity. Statistical approximations are then used to calculate the expected duration for each activity and for the entire project.

"Project": A project is any endeavor for which an objective, a beginning, and an end can be clearly defined.

"Activity": An activity is an element of work, a task that constitutes part of a project. When all the activities are complete, the project is said to be complete. It should be noted that an activity can be defined to any degree of detail, and a project can be defined as having many or few activities depending on the amount of detail in each activity. An activity has a duration that is determined by its scope of work, the labor and equipment resources that will be applied, and the basic conditions under which the work will be done. An activity also has an identifiable start and finish. Preferably, it also can be assigned to a specific individual, group, or company. A critical activity is an activity on the critical path.

"Event": An event is a point in the project that has special significance. A "milestone" is a commonly used term for event. All projects have at least two events: the start and the finish. Other events, or milestones, typically include the approval of funds, the delivery of critical

equipment, or the award of an important contract. Milestones are a useful way of keeping track of many small projects as they give a quick indication of whether or not they are on schedule. Unlike an activity, an event (or milestone) has a duration of zero and no resources or costs are applied to it.

"*Network Plan*": A network plan is a graphical representation of the dependencies between the activities in a project. The laws of nature dictate that some activities, such as preparation of the foundation, must precede others, such as erection of structural steel. Standard engineering and construction practice also dictates that certain sequences be followed for the installation of piping, electrical facilities, etc. These conditions that determine when an activity can start or finish are "constraints" or "dependencies."

"*Network Notation*": Networks can be drawn using two types of notation:

"*Arrow*" *diagrams*: These define the start (or "I" node) as well as the finish (or "J" node) of each activity. This type of notation is illustrated in Figure 3.1, and is also referred to as "I-J," "bubble," or the "Arrow Diagramming Method" (ADM). An arrow diagram would quickly become hard to read were it not for the use of "dummy" activities. The dummy simply indicates that two separate nodes are actually the same: that the two activities, whose finishes are marked by the nodes at both ends of the dummy, must both be completed before the next activity can start.

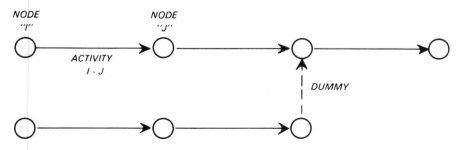

Figure 3.1   Arrow Diagramming Method (ADM). Each activity is defined by its start node "I" and its finish node "J."

"Precedence" diagrams: These define each activity in terms of the activities which must be completed before that activity can start. A precedence network is shown in Figure 3.2. It can be seen that a precedence network does not use nodes or dummies. Although the illustration shows the finish-to-start relationship, it is also possible to diagram start-to-start and finish-to-finish relationships. Precedence notation is called "Precedence Diagramming Method" (PDM). The activities that preceed the activity in question are called "predecessors," and those that come after it are "successors".

In comparing arrow and precedence notation, we generally find that precedence is easier to understand and use for those who are not trained planners. However, since one need only specify two node numbers to identify any activity, no matter how many predecessors it has, arrow format can be simpler for large, complex networks. Many computer programs available today can use either notation, although most PC programs use precedence.

"Critical Path": After a network has been drawn—showing the activities, their durations, and constraints—it is evident that there is a certain sequence of activities (or "path through the network") whose total duration exceeds that of all other paths. That is, the sum of the durations of each of the activities on this path will be equal to the total time required to complete the project. *If any activity on this path should slip, the completion date of the project will slip a like amount.* This path through the network is the "critical path," and all the activities on that path are "critical activities." Critical path is a very useful concept for small projects.

"Float": For those activities not on the critical path, it is evident that some slippage in either the start or completion can be tolerated with-

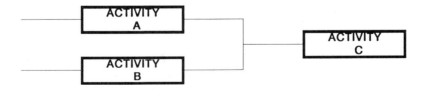

Figure 3.2 Precedence Diagramming Method (PDM). Each activity is defined by specifying those activities that must precede it (e.g., activity C cannot start until activities A and B are complete).

out causing any slippage in project completion. This amount of slippage is referred to as float (also called "slack"). There are two types of float:

"Total float" (*TF*): This is the amount of time that completion of an activity can slip without affecting the completion date of the project. Total float is the best understood and most commonly used float calculation. The word "float" often denotes total float. Note that the use of total float may not delay project completion, but it may delay successor activities thereby increasing the likelihood that project completion will be delayed.

"Free float" (*FF*): This is the amount of time that completion of an activity can slip without it affecting another activity. For example, an activity can slip by less time than its TF, and still affect the starting date of its successor. FF provides the best indication of the amount of slippage that can be easily tolerated (See "Misconception: Float is a Form of Contingency," later in this chapter). It is also useful to indicate which activities are likely to become critical. Most project management programs will calculate and display FF.

Activities on the critical path have, of course, no float.

"Time Analysis": Float times, and the critical path, are calculated by a procedure known as time analysis. Time analysis consists of a "forward pass" and a "backward pass" to determine critical path, float times and the "early-start," "late-start," "early-finish," and "late-finish" times for each activity.

"Forward pass": Beginning at the "start" of the network, durations are added to determine the early-start and early-finish times for each activity.

"Backward pass": Beginning at the "finish" of the network, durations are subtracted to determine the late-start and late-finish times for each activity.

"Early start time" (*ES*): This is the earliest time at which an activity can begin. The early-start time will be identical to the early-finish time of the preceeding activity.

"Early-finish time" (*EF*): EF is the earliest time at which an activity can be completed. It is the ES plus the activity duration.

"Late-start time" (*LS*): This is the latest time at which an activity can begin without causing a delay in project completion.

"Late-finish time" (*LF*): LF is the latest time at which an activity can be completed without causing a delay in project completion. It is the

LS plus the activity duration and is identical to the LS of the succeeding activity.

From the above it can be seen that, for any given activity:

$$TF = LS - ES = LF - EF$$

when $ES = LS$, $EF = LF$, $TF = 0$, the activity is critical.

"Hammock": In order to simplify a network for those with no need for details, a hammock is often used. A hammock, illustrated in Figure 3.3, allows us to represent several activities in the network with one activity. The detailed activities are often called a "subproject." The concept of the hammock is very important to multiple small-project planning where we need to avoid the use of detailed reporting, while preserving the level of detail necessary for effective control. Hammocks make it possible to display a project at various levels of detail, to suit the point of view of the user. A management network, often referred to as a "Level 1," provides the summary-level view a manager requires. By contrast, the drawing-office supervisor might be working with a detailed plan that shows all the activities associated with each drawing.

This "network hierarchy," made possible by hammocking, is also a useful tool for managing the contractor or subcontractor. For this purpose, a "subproject" is prepared to represent the work described by the single activity on the higher-level network. The subproject allows detailed control procedures to be implemented without causing that extra detail

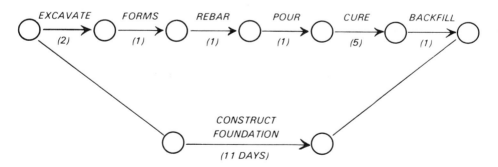

Figure 3.3   A network hammock. The single hammock activity, "construct foundation," represents the scope of work and total duration of the six detailed activities.

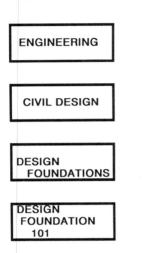

ENGINEERING	LEVEL 1 PLAN
	(MANAGEMENT LEVEL)

CIVIL DESIGN — LEVEL 2 PLAN (PROJECT MANAGEMENT LEVEL)

DESIGN FOUNDATIONS — LEVEL 3 PLAN (SUPERVISION LEVEL)

DESIGN FOUNDATION 101 — LEVEL 4 PLAN (DETAILED SUPERVISION LEVEL)

Figure 3.4 Use of hammocks to establish the network hierarchy. This example shows the activities for an engineering project.

to be visible at the management level. The use of subprojects is shown in Figure 3.4. Most PC-based project management software provides this subnetwork capability (see Chapt. 7 and 15).

"*Barchart*": A barchart is a tool for scheduling, that is, for representing the dates during which each activity will take place. Unlike a network, a barchart does not show the constraints between activities. Barcharts are often referred to as "Gantt Charts."

FUNDAMENTALS OF CPM (THE CRITICAL PATH METHOD)

The techniques for planning and scheduling a small project using the Critical Path Method are described in this section. The basic technique consists of the following simple steps:

1. Define the scope of work of the project
2. State the assumptions on which the plan and schedule will be based
3. Break the project down into activities
4. Define duration, resources and costs for each activity
5. Define dependencies and draw network
6. Calculate critical path and float

7. Define the calendar(s)
8. Schedule activities
9. Optimize plan and schedule by iterations of steps 1-8

Step 1: Define the Scope of Work of the Project

As most contractors know, the failure to clearly define and agree on the scope of work prior to starting a project is one of the most frequent causes of cost overruns and schedule delays. There are a number of reasons for this, such as the common perception that time is short and everyone understands what has to be done anyway. Another is the feeling that, if the full scope of work is revealed, the project might not be approved. Even when design specifications are prepared in detail, experience shows that we often encounter many unforeseen problems on a small project. Scope assumptions can help by stating explicitly what is included and what is excluded from the plan.

In preparing a plan for the small project, we generally begin with a technical description of the project: marked-up drawings, design specifications, etc. Our first step in defining our scope of work for planning purposes is, therefore, to assure ourselves that the design basis for our plan is reasonably complete. This might entail a trip to the field (to clarify both the design and construction work required), or a review of similar past projects. We should be able to reach agreement on the documented technical scope. Given an agreed and understood technical scope, are we now ready to begin planning? Unfortunately, we are not. Like so many other aspects of small projects, a special problem arises: the problem of "who does what." The point at which the project is handed over to the project leader and the point at which it's completed and handed back to the client must be defined. Many project engineers have been heard to cry, "But I was told that the site clearance would already be done before we arrived!", or, "They said that all the buildings and utilities for construction would be provided!" Of course, when these things have not been done, we are left to do them ourselves and hope we can get a scope change approved to cover them.

When is a project "complete"? Does a project's scope include startup and commissioning of all systems? Does it include final revisions to all drawings? Does it include all "punchlist" items? The answer is, "it all depends." Some projects include these items, and some don't.

If a clear definition of the technical and planning scope of work cannot be easily obtained, the project leader should clearly document the scope assumptions, and thereby set the basis for discussing changes if the assumptions turn out to be incorrect.

Step 2: State the Assumptions on Which the Plan and Schedule are Based

A plan or schedule is only as good as the assumptions on which it is based. Some of the assumptions which have a marked influence are:

Work schedule: number of shifts per day, number of working days per week, use of overtime, holidays

Work content: conditions at start, definition of project completion, amount of work which is currently undefined

Quality requirements: normal practice, above-normal practice, sacrifice of quality for work speed?

Cost vs. schedule priorities: OK to spend money to expedite schedule?, is minimum cost the primary objective?

Availability of resources: skilled labor, critical materials, special equipment

Efficiency of management decision-making: can critical decisions be made quickly?

Access to work site

Contractor performance: productivity, mobilization time, quickness of response

Market conditions

Labor conditions

Supplier performance: time to bid, delivery time, quality vs. the need to fix in field

Weather conditions: time of year, and location where outdoor work is to be done

Assumptions should be made to reflect the most probable situation, not the most optimistic. When a file of similar past projects is available, the assumptions should reflect the average past experience. If one or more key assumptions are open to question, a plan and schedule can be prepared for each case, so that the sensitivity can be seen.

Step 3: Break the Project Down into Activities

With the scope of work clearly defined, we are now able to define the specific activities required to accomplish that work. As noted previously, an activity can be defined to any level of detail, depending on the need for control. We can say that the project leader's level of detail is Level 2; that is, more detailed than that used by a manager but less detailed than that used by those who are directly responsible for only one part of the project. The outlining feature found in some project management programs is very useful here.

What determines the level of detail of an activity? The level of detail should be appropriate to the amount of planning and control required. For example, we might have a project in which the work is to be done on a "cost-plus" basis. Clearly, this is a case in which close monitoring of workhours and progress will be required. In this case, our project leader's network will contain all the activities describing major aspects of the work. Suppose that same engineering work was to be done under a lump-sum contract? Although still responsible for the contractor's performance, our project engineer's network will probably show mostly those activities relating to technical quality, as less control of workhours is required.

The activities should be consistent with the contracting plan, and defined in sufficient detail to provide a basis for resource planning, cost estimating, progress measurement, and control. Most small projects can accomplish this with networks of 50 activities or less. Each activity should be significant, cover an identifiable scope of work, and have a clearly defined start and finish. In the case where the detail required for control exceeds the detail of the Level 2 network, we can use subnetworks to describe activities for control as shown on Figure 3.4. It is important that activities be assignable to a specific individual, group, or organization.

Step 4: Define Duration, Resources, and Costs for Each Activity

The expected duration for an activity is that elapsed time that we expect will be required to complete the scope of work described by the activity, with the resources we expect to be available, if our basic assumptions are correct. By definition, the probability is 50% that the actual duration will be equal to or less than the expected duration (see Figure 3.5).

This definition, given to us by statistical methods and illustrated by PERT theory, points out an important concept that is often overlooked, even by experienced planners. If we calculate the most-likely activity duration, that is, the duration that will occur more frequently than any other duration, we find that the probability of that duration actually being achieved is generally less than 50%. This is due to the fact that the most optimistic duration is, typically, around 70% of the most likely (you can

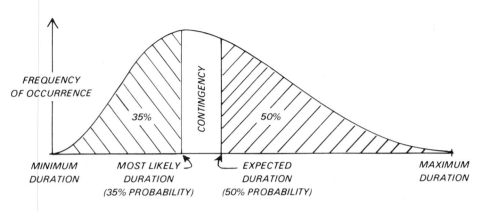

Figure 3.5   Probability function for activity durations. The area under the curve indicates the probability that the actual duration will be equal to, or less than, a given duration. The difference between the expected and most likely durations can be considered to be schedule contingency.

only accelerate the schedule just so much), whereas the most pessimistic duration can be 300% or more of the most likely. This illustrates the principle of "skewness," and is as good an illustration of Murphy's Law as we are likely to find. Because, in general, there is a limit on how much we can improve things, while there often seems to be no limit at all of how much things can go wrong, the expected (or 50%-probability) duration, will be larger than the duration we say is most likely. The consequences of confusing the most likely and expected values will be explored later in this chapter (see "Some Popular Misconceptions,") and also in Chapter 6.

The best method for establishing the most likely duration is the use of historical data. Most small projects are similar to other small projects by the same company, and it is therefore possible to develop data records which show, on average, how long certain tasks take. If such a database is available (see Chapters 9 and 14 for details), we need only examine the project at hand for differences with the average past project, and make adjustments accordingly. If such information is not available in-house, there are a number of industry services that provide such data. Before using such a service, the database should be checked against some actual data, and factors defined to be applied to the results from the service. Finally, there are numerous judgements and rules of thumb that can be used. The network can, of course, be used to test the sensitivity of the schedule to the durations assigned to various activities. Given a good estimate of the most likely duration or cost, the PERT method can be used to calculate the expected duration.

Step 5: Define Dependencies and Draw the Network Plan

Now that we know the activities that constitute the project we are ready to plan the sequence in which those activities will be performed, and present that plan in the form of a network. To establish our network logic, we must identify activity dependencies; that is, those activities that cannot start until one or more other activities are complete. We can group dependencies into two categories:

1. *Mandatory.*   Those dependencies that cannot be violated; for example, formwork and reinforcing steel must be installed before concrete can be poured. These are sometimes called "hard logic" dependencies.
2. *Arbitrary.*   These dependencies represent good practice and reflect the way in which we intend to do the work, but they can be changed. For example, though a pump should be installed before its piping, if delivery of the pump is delayed, the piping can be installed first and final corrections made later. This is referred to as "soft logic."

As the dependencies for each activity are identified (in terms of those activities that precede it), it is helpful to note which category the constraints fall into, so that the logic can be changed, if necessary, to optimize the network. The network should be drawn assuming that no constraints are imposed by limitations in resource availability.

A network can now be drawn. When the network is complete, it is useful to check for "loops," i.e., errors in logic that create a circular flow instead of a left-to-right linear flow. Another potential error to check for is "dangles," i.e., errors in logic which leave an activity with no connection to another activity.

When drawing the network it is also useful to identify the milestones that will be used later as a quick reference on schedule status. Some companies have found it helpful to define standard networks for groups of activities or for entire projects that are often repeated. Maintenance projects, some engineering projects, product development and certain types of light construction lend themselves to this approach. Standard networks reduce planning time, increase consistency, and help assure that all necessary activities are included in the plan. The network plan for a specific project can be derived simply by changing and adding activities or constraints as necessary. It should be noted that the network can be drawn easily by most PC project management programs.

Step 6: Calculating Critical Path and Float

In this step, a time analysis is performed on the network using the following procedure.

Starting at the "start" event, a *forward pass* is made. For each activity, in turn, we write down the early-start time using the notation shown in Figure 3.6. For the first activity (or activities) the earliest possible time

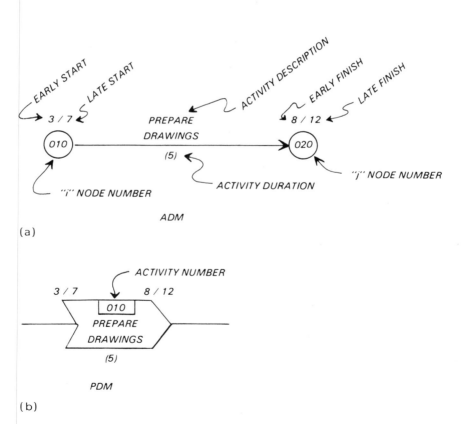

(a)

(b)

Figure 3.6 Network notations: (a) arrow (ADM). (b) precedence (PDM).

that the activity can be started is at day #1. If the duration of the first activity is 2 days, then the earliest time at which that activity can be completed is at an elapsed time of 3 days (see Fig. 3.7a). This is also the earliest time that the succeeding activity can start. The forward pass procedes through the network in this manner. It can be seen from Figure 3.7a that the early-start time of an activity with several predecessors will be the latest of the early-finish times among the proceeding activities. When the finish event is reached, the network should show the elapsed time to each activity's early-start and early-finish.

In order to calculate the critical path and the float of the noncritical

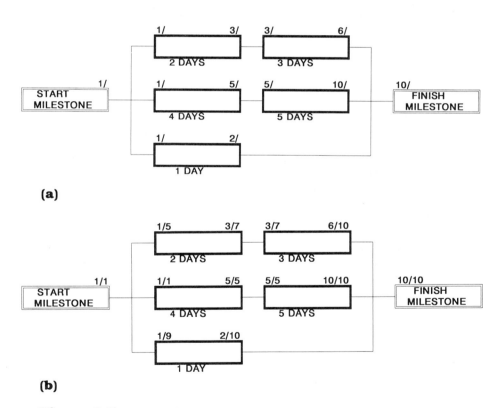

Figure 3.7   (a) Forward pass: calculating early-start and early-finish times. (b) Backward pass: calculating late-start and late-finish times.

activities, a *backward pass* is made. Starting at the finish event, the latest finish of the last activity can be seen in Figure 3.7b to be equal to the early finish. By subtracting the duration of the last activity from the late-finish time, the latest-possible start time is identified. The backward pass continues in this manner. When an activity has several predecessors, its late-start time becomes the earliest of the late finishes of all the predecessors. When the backward pass is complete, the late-start and late-finish times are shown for each activity.

The critical path is now evident from inspection of the network. Those activities for which the early start/late start and early finish/late finish times are identical are critical. There should be a path through the network made up of critical activities; if not, there is an error.

For those activities for which there is a difference between early start/late start and early finish/late finish, that difference is equal to the total float.

It should be noted that there may be more than one critical path.

Step 7: Define Calendars

The "calendar" used in project management software is the mechanism by which planning information is converted to a schedule. The calendar defines working and non-working days and hours.

For example, a non-working day is a weekend or holiday, nonworking hours may be from 5:00 pm to 8:00 am. Given an activity with a certain duration and set of dependencies, the program will consult the calendar to determine the actual start and finish dates.

Therefore, calendars are a very powerful and important function in project management software. Many programs offer multiple calendars, such as one calendar per resource. This allows different activities to be scheduled according to the appropriate calendar. For example, activities related to product design might be scheduled according to an office calendar of 40 hours/week, whereas activities related to product testing might be scheduled according to a laboratory calendar of 7 days/week, 24 hours/day.

Step 8: Schedule Activities

With the time analysis complete, the project is now ready for scheduling, that is, defining the date or time at which each activity will begin and end. Given a start date, the dates for early/late start/finish can be easily determined, and the project-completion date defined.

A barchart is the most common way of representing the schedule. For those activities with float, the barchart can show them scheduled to start at the early-start date (leaving the float unused), the late-start date (all float used prior to the start), or some time in between. Barcharts often indicate the float available by using a different shade for the float time (see Figure 3.8). It is important to distinguish between the uses of barcharts and networks

Networks	Barcharts
Are used for planning	Are used for scheduling
Show dependencies	Do not show dependencies
Show durations and elapsed time	Show durations and dates
Change infrequently during project	Change frequently

A barchart should always be based on network logic, even if that logic is not shown. Many companies have found that a format combining the best features of the barchart and the network can be useful. Some of these formats are:

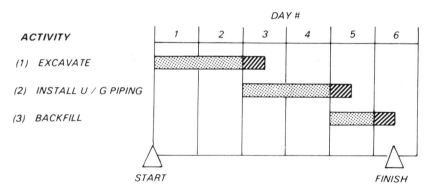

Figure 3.8   Bar chart showing total float. Each activity in the sequence has one-half day total float. The bar chart shows early start for all activities, but does not show dependencies.

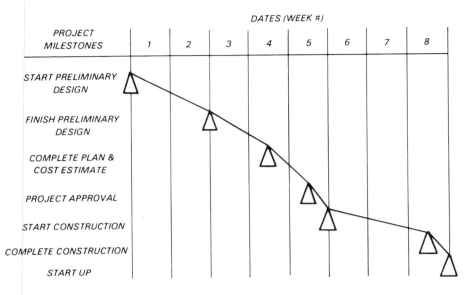

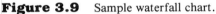

Figure 3.9   Sample waterfall chart.

A network plotted on a time scale
A barchart showing constraints
"Waterfall" or "cascade" chart showing milestones and dates

An example of the waterfall chart is shown in Figure 3.9.

Step 9: Optimize the Plan and Schedule

It is evident from the procedures described above that planning is an iterative process. The assumptions on which the plan and schedule are based
are made, in general, without a good idea of the actual impact they will
have on the project. For example, it may have been assumed that certain
activities will be performed on a crash basis—until the time analysis reveals that those activities have sufficient float to make that unnecessary.
Or, our logic diagram may have resulted in a critical path that included
activities that need not be critical, so a simple change in the logic is all
that is required to improve the schedule. The scheduled completion date
is often found to be unacceptably late. The network logic and the basic
assumptions must then be examined and changes made accordingly. As

we change start dates, logic, or durations—and iterate in order to optimize the schedule—we are, in effect, using our project model for decision-
making.

A good procedure when using project management software is to
build the project model as follows:

Step 1: Fix the date of the Start milestone: This locks the beginning of
the schedule while preserving the flexibility required for modeling.

Step 2: Let all activities be "ASAP": This provides the Early Start schedule for each activity.

Step 3: Compare the program-generated dates with known deadlines:
The calculated dates, although consistent with the dependencies,
durations and calendars you have input, may not be acceptable.

Step 4: Edit the plan until an acceptable schedule is reached: This may
involve adding resources to shorten durations of critical activities, paying premiums to expedite materials, changing the network logic to shorten the critical path, changing the calendar,
breaking up activities to enable the critical path to be shortened

COMPLEX DEPENDENCIES

An important tool in optimizing a plan is the use of "complex dependencies" i.e., dependencies that allow activities to be overlapped.

Consider the traditional Finish-to-Start dependency illustrated by
Figure 3.2. It assumes that all the work in the predecessor activity must
be complete before any work at all can start on the successor. When we
think about it, we can see that this is a very conservative assumption. In
many cases, we can at least do some work on the successor task before
the predecessor is 100% complete.

The problem with excessive use of the Finish-to-Start dependency is
that it is apt to produce a schedule that is too long. Alternative types of
dependencies, which permit overlapping of activities, are very useful.

The Start-Start Dependency

Consider the activities shown in Figure 3.10. We all know that Procurement is dependent on having some design work done. But must all the
design work be done before procurement can start? Certainly not! What
determines the start of the Procurement activity is the start of Design plus
a "lag" of several days. We therefore have a Start-Start dependency plus
a lag of +3 days.

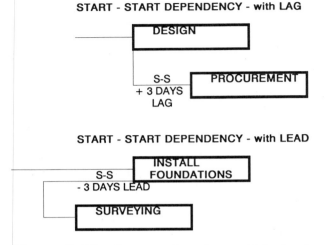

Figure 3.10   Start-start dependency, with lead or lag, allows overlapping of activities constrained by their start date.

In this case we are instructing our project management software to schedule the start of the predecessor, then schedule the start of the successor three days later.

Note that a "lag" refers to a delay in the start of the successor. Therefore, it is conceivable that we could have a "lead," that is, a successor whose start depends on the start of the predecessor but that actually must begin before the predecessor. Such a relationship is illustrated in Figure 3.10. We wish to start surveying just before the installation of foundations is to start—if the foundation work is postponed, so will the surveying. This relationship is established by a Start-Start dependency plus a lead time of − 3 days.

The Finish-Finish Dependency

Consider the activities shown in Figure 3.11. We can see that much of the electrical work (e.g., running cable and conduit) can be done independently of equipment setting. But we also know that the electrical work cannot be completed until 3 days after the last equipment item has been installed. This extra time is required to make final connections to the equipment.

FINISH - FINISH DEPENDENCY - with LAG

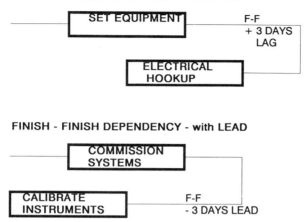

FINISH - FINISH DEPENDENCY - with LEAD

Figure 3.11 Finish-finish dependency, with lead or lag, allows overlapping of activities constrained by their finish date.

This relationship can be easily represented by a Finish-Finish dependency with a lag of 3 days. In this case we are instructing our project management software to schedule the finish of the predecessor, then schedule the finish of the successor three days later.

Figure 3.11 also shows an example of the Finish-Finish dependency with a "lead." In this example we can be calibrating instruments independently of commissioning, but since we need all instruments operating for at least the last three days, the calibration must be finished three days before the commissioning.

The Finish-Start Dependency

Figure 3.12 illustrates some of other uses of the Finish-Start dependency. There may be some cases where we wish to show that an activity cannot start until some time after its predecessor has finished. For example, this lag time might represent the time we wait for government approval.

The Finish-Start dependency can also be used to overlap activities when a "lead" is provided. This is a useful technique when the duration of the predecessor is uncertain. For example, in Fig. 3.12 we see some

FINISH - START DEPENDENCY - with LAG

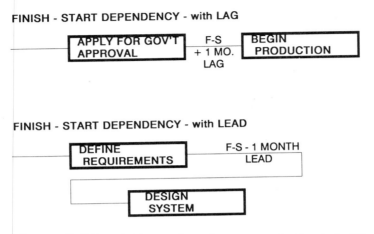

FINISH - START DEPENDENCY - with LEAD

Figure 3.12 Finish-start dependency constrains the start of the successor by the finish of the predecessor. Use of lead allows activities to overlap.

activities taken from a computer system design project. "Define Requirements" involved talking with many people and its duration is uncertain. But the programmers will know when they are about 1 month from completion of "Define Requirements" and that is when they will start "Design System."

WORK BREAKDOWN STRUCTURE

A planning technique that has been gaining acceptance in the multiple small-project environment is the Work Breakdown Structure (WBS). WBS is really a first step in planning, where we break a project down into manageable components in order to make planning easier.

Note that the project is broken down into physical systems or components, and then activities are defined for each one. This is illustrated by Figure 3.13a.

Although WBS was originally considered to be a planning tool for large or complex projects, it has become useful for multiple small-projects because it provides a uniform framework that can be used consistently on all small projects.

An example of the use of a WBS code to identify small project activities is shown on Figure 3.13b. Note how all projects can fit the structure. Imagine now that we had a multiple-project plan with 50 projects on it.

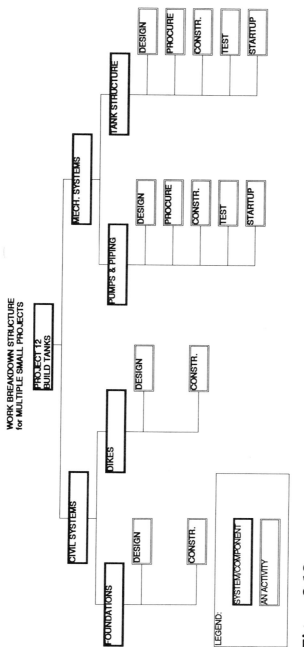

WORK BREAKDOWN STRUCTURE
for MULTIPLE SMALL PROJECTS

PROJECT 12
BUILD TANKS

CIVIL SYSTEMS

MECH. SYSTEMS

FOUNDATIONS
DIKES

DESIGN
CONSTR.

DESIGN
CONSTR.

PUMPS & PIPING

TANK STRUCTURE

DESIGN
PROCURE
CONSTR.
TEST
STARTUP

DESIGN
PROCURE
CONSTR.
TEST
STARTUP

LEGEND:

SYSTEM/COMPONENT

AN ACTIVITY

Figure 3.13a   Work breakdown structure organizes each project by physical components.

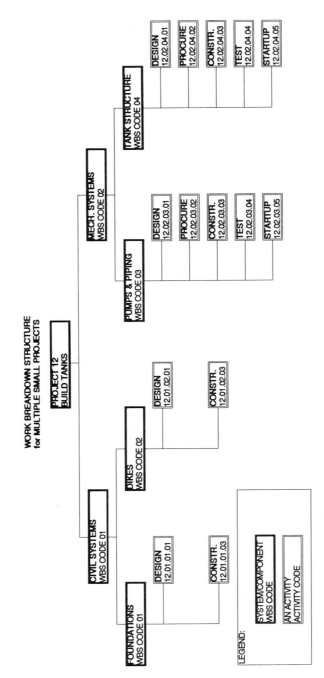

Figure 3.13b The WBS code identifies the area system or component to which an activity belongs.

Suppose we wanted to extract certain information. We could use the DOS wildcard symbols (i.e., * and ?) together with the selection or "filtering" capabilities of our project management software to do that.

For example to extract all the data pertaining to civil work on project 12, we would set up a filtering criterion as follows:

WBS code = 1201*

To select the civil work on all projects we would use:

WBS code = ??01*

To select all information pertaining to the pumps and piping on all 50 projects we would use:

WBS code = ??0203*

Many PC-based project management programs offer a number of features to make implementation of WBS easy and effective.

Planning and scheduling, in spite of all the arithmetic involved, is still something of an art. Much of the artistry takes place during the iterations of the network, as the variables are changed until a satisfactory plan and schedule are achieved.

SOME POPULAR MISCONCEPTIONS

As CPM planning has gained in use, a number of misconceptions have grown around it. For the technique to be effective in the small-project environment, these misconceptions must be avoided.

Misconception: Float Is a Form of Contingency

Because TF shows the time that an activity can slip without affecting the end date, it is often thought that it makes no difference whether the float is "used" or not. The float is therefore thought to be a form of contingency, a "slush-fund" of time to take care of unforeseen problems. This is a dangerous misconception that can lead to large schedule slippage.

Let us consider a series of activities each of which has float (see Figure 3.14). Our barchart at the start of the project shows each of the activities with a TF of ½ day. Suppose that the project is now underway, and the first activity has been completed at the late-finish date, thereby using all of its TF. It is immediately apparent that the remaining activities in the series must all now start on their late-start dates. They have, therefore, lost all their float and are critical. So, float is too easily lost to be considered contingency. Another reason that float cannot be considered

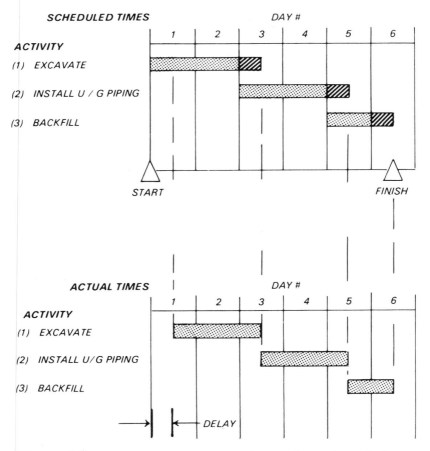

Figure 14   Effect of losing total float. If excavation begins on its late-start time, thereby "using" its float, it and all succeeding activities on that path become critical.

contingency is that it provides the flexibility needed to compensate for lack of resources (see Chapter 4). Finally, the variations covered by contingency (e.g., design changes) may well affect critical activities and, the more critical activities, the greater is our risk of not completing the project on time!

The concept of FF is a truer representation of the amount of slippage an activity can tolerate before it affects the schedule. In other words, the

FF can be used up without affecting any other activity. An example is shown in Figure 3.15. Free float is the difference between the early finish of an activity and the earliest of the early-start times for its successors. It is particularly useful when someone else has the responsibility for the succeeding activity. When we use our TF, he is affected as his float is lost, but when we use our FF, he doesn't know the difference.

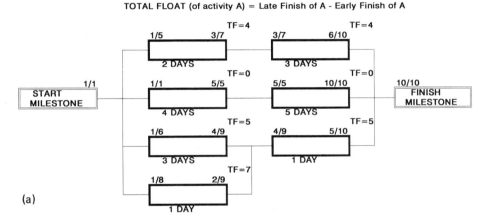

(a)

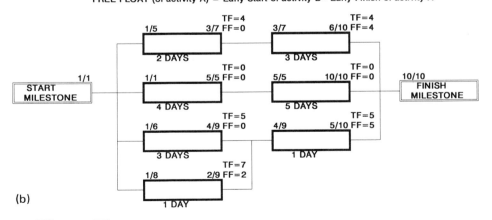

(b)

Figure 15 Two types of float. Figure 3.15a shows total float that affects the end date and other activities. Figure 3.15b shows free float is the slippage in an activity that can occur before any other activity is affected.

Misconception: Target Schedules Help Assure On-Time Completion

Many project engineers and managers believe that a tight schedule helps to put pressure on those doing the work, thereby helping to assure on-schedule completion. One reason why this seldom works is the fact that this approach has been used for so long that it hardly fools anyone anymore. Most contractors know that the client is apt to have a budget and schedule that contains some contingency, and to conceal this fact only leads to a loss of credibility. Another reason is due to a simple but often overlooked principle of probability. If we accept that the probability of success decreases as the schedule becomes increasingly "tight," and that costs increase as we pursue the tighter schedules, it follows that we can easily fall into the trap of spending large sums of money in the pursuit of schedule goals with a low chance of success. Thus the enhanced profitability that the shorter schedule is meant to achieve will not be realized.

To avoid this problem, the project leader should be sure that the durations assigned to each activity have an equal probability of being exceeded or underrun. This can be calculated with the PERT method described in Chapter 6.

Misconception: The Network and Barchart Form a Complete Project Plan

Although the complete network and barchart would seem to be more than enough to assure that the schedule is realistic, there are two important factors that are often overlooked: (1) the plan for the application of material and labor resources, and (2) the contracting approach that will be taken to assure that those resources are provided in a cost-effective way.

The resources that are applied to the project determine, more than any other single factor, how long the project will take. Work cannot be accomplished without the people, the machinery, and the materials necessary to do it. Yet, in spite of the importance of resource analysis to identify resource requirements and assure that the manpower, machinery, and materials are available when needed, this technique is often underutilized. Resource analysis is particularly vital for small projects because of the small project's need to compete for resources, as well as the lack of alternative ways to make progress if essential resources are not available when required. For example, if a small project involves replacement of a heat exchanger, and that heat exchanger is not delivered on time, that project will certainly be delayed. By comparison, a large project will have other materials on-site, so progress can be made. Because of its importance to small projects, resource analysis will be described fully in Chapter 4.

The contracting plan describes how contracts will be arranged to assure timely provision of the resources and services required. Many project leaders overlook the extent to which contracting options are available, and the extent to which the selection of a contracting approach can affect the schedule. This is discussed in the next section.

CONTRACT PLANNING

In many small projects, much of the work is done by contractors or subcontractors. The approach taken for bidding, negotiating, and managing these contracts can have a major impact on the project's cost and schedule, as well as on the quality of the work done. Many small projects are conducted without proper attention to defining a contracting strategy that is right for the project. The optimum blend (referred to in Chapter 2) of cost, schedule, and quality will be different for different projects, and the contract plan should be designed to reflect these basic needs.

The contract plan for a small project may be nothing more than a thought process that considers the special needs of the project at hand and the different contracting approaches that are available, and makes a decision on that basis. The network plan and schedule can also be used to evaluate different contract-plan scenarios. For example, one plan might cost more but result in some schedule improvement, and the project model could be used to quantify the schedule benefits that could be expected.

In this and subsequent sections we will refer to the relationship between "client" and "contractor." It should be recognized that every organization is both a client and a contractor: the project leader is, in effect, a contractor to the operating plant that is often referred to as a client. Engineering and construction contractors are, of course, the client to the firms to which they issue subcontracts. So it is important to consider the client/contractor relationship with an appreciation of both points of view.

Factors That Determine the Contracting Plan

Some of the factors that affect the selection of a contracting plan are:

Market conditions: Are the contractors looking for work or do they have all they can handle?

Project size: Is this project bigger or smaller than most other projects the contractor has done before?

Contractor capability: Has the contractor demonstrated the ability to handle the size, complexity, or technical discipline of the type of work involved?

Schedule: Is the schedule particularly important on this project? Is time available to pursue a contracting approach that requires more time for negotiation? Is cost more important than schedule?

Technical security: Does the project involve proprietary technology? (many research projects fall into this category)

Financing: Are there any special financing considerations? Is the contractor expected to do the work and be paid later, thereby financing the project himself? Should he be? Can money be saved if the client finances the project at the lower interest rates he can probably secure?

Risk: Are there any special technical, cost, or schedule risks associated with the project? For example, are the penalties for late completion greater than normal? What is the impact of a cost overrun? Is there any uncertainty about the technical basis, i.e., has the design ever been built before?

Company image: Are there any limitations on the selection of contractors imposed by the desire to project a certain image? For example, many companies prefer to work with local or domestic contractors and suppliers to foster goodwill, even when better price and performance could be obtained elsewhere.

The answers to these questions will be different for most projects, and the contracting plan should be designed accordingly.

Basic Types of Contracts

There are three types of contracts: lump-sum contracts, reimbursable contracts, and contracts that are combinations thereof. These are summarized in Figure 3.16.

Lump-Sum Contracts

Lump-sum contracts (also referred to as "fixed-price" or "hard-money" contracts) set a price for a defined scope of work. A pricing structure is also agreed upon for adjusting the lump sum for changes made by the client. The lump sum includes the contractor's direct costs, overhead, and fee. The lump-sum price may be paid in full upon completion of the work, but it is most often paid in increments according to progress.

ATTRIBUTE	LUMP SUM	REIMBURSABLE	REIMBURSABLE W/FIXED FEE	UNIT PRICE
PRICING	HIGHLY COMPETITIVE	COMPETITIVE	COMPETITIVE	COMPETITIVE
CONTRACTING TIME REQUIRED	LONG	SHORT	SHORT	MEDIUM
SCOPE OF WORK DEFINITION	DETAILED DEFINITION, FIXED SCOPE	GENERAL DEFINITION, VARIABLE SCOPE	GENERAL DEFINITION, VARIABLE SCOPE	SEMI-DETAILED DEFINITION, VARIABLE SCOPE
CLIENT RISK OF COST OVERRUN	LOWER	HIGHER	MODERATE	MODERATE
POTENTIAL FOR CLAIMS	HIGH	LOW	LOW	MODERATE
MARKET CONDITIONS REQUIRED	COMPETITIVE	NONE	NONE	MODERATELY COMPETITIVE
NEGOTIATION EFFORT	HIGH	LOW	MODERATE	MODERATE
CONTROL AND ADMINISTRATIVE EFFORT	LOW	HIGH	MODERATE/HIGH	MODERATE

Figure 3.16   Summary of contract types.

Advantages of lump sum   The theoretical advantages of the lump-sum contract are:

Competitive pricing: In market conditions characterized by good competition, the lump-sum approach can result in the lowest-possible price being obtained, as contractors may be motivated to cut profit margins while maximizing performance. With a lump-sum contract, the contractor has an incentive to perform in a cost-effective manner.

Reduced risk to the client: Because the final cost is known in advance, the risk of cost overrun is theoretically eliminated. Any overruns should be to the contractor's account.

Reduced control required by the client: Since the price is fixed, no cost control is necessary

Because of these attractive advantages, many project leaders and managers believe that contracts should be lump sum whenever possible.

This is a very dangerous policy that has resulted in a lot of unnecessary overruns and many unpleasant negotiations. In fact, contracts should be lump sum whenever it fits the project's design and schedule requirements and the prevailing market conditions. There are, however, some serious disadvantages to lump-sum contracting.

Disadvantages of lump sum   The disadvantages of the lump-sum contract should be carefully considered before a decision is made:

1. The scope of work must be well defined. If any of the lump-sum advantages are to be realized, the scope of work and price that is initially agreed upon cannot be significantly changed. When changes to the scope do occur, the contractor is in a noncompetitive situation and can be expected to take full advantage of it. This is a particular problem for many types of small projects, such as revamps, turnarounds, and major maintenance in which the scope of work cannot be known with any certainty until the project is well along.
2. The schedule must provide adequate time for the client to define the scope of work to a sufficient level of detail that it can be the basis of a firm bid with low probability of significant changes.
3. The schedule must provide sufficient time for the contractors to prepare bids, and for the client's bid review and selection process. On small projects, this amount of time may be a disproportionate part of the overall schedule.
4. The client's risk is only partially reduced, due to the understandable reluctance of the contractor to let one contract put him out of business. Should changes, unforeseen circumstances, or client-contractor communication problems reach a high enough level the contractor can, quite justifiably, be expected to submit a claim. Once the work is underway, the contractor has the upper hand in any negotiation because the cost to the client of the contractor's failure to perform probably far exceeds the value of the contract. In situations where there is significant risk, the contractor's bids may include contingencies to such an extent that any cost savings from the lump-sum approach are lost.
5. The client is still obliged to monitor and control the progress and technical quality of the work, as well as be alert for potential claims. Since the price has been fixed, quality control is especially important to assure that no corners have been cut to save cost.
6. The willingness of bidders to invest the time and effort required to prepare a lump-sum is very much a function of market conditions. If

the contractors are busy on profitable work, they may either decline the opportunity to bid or simply submit a high bid. So, successful lump-sum contracting requires favorable market conditions in which several qualified bidders are willing and able to prepare bids.

In summary, many companies have made the mistake of rushing into a lump-sum contract with an incompletely defined scope of work, only to find that the subsequent scope changes resulted in substantial cost overruns and schedule delays, as well as costly negotiations, all of which could have been avoided. Lump-sum contracts are indeed a way to save money and extract the best performance, but these results only occur after careful planning and execution of a lump-sum contracting plan under the right conditions.

Reimbursable Contracts

Reimbursable contracts (also called "cost-plus" or "time and material" contracts) are those that reimburse all of the contractor's direct costs fully, in addition to which he is paid for overhead and fee. In its most common form, this type of contract provides an all-in rate for each hour of services provided, as well as a markup on materials and services procured on behalf of the client.

Advantages of reimbursable contracts

Shorter contracting time. Because it is only necessary to define the scope of work for a reimbursable contract in general terms, the time required for the client to define the scope and for the contractor to prepare his bid is greatly reduced. On a small project, the schedule may be short enough that the time available for contracting will only permit a reimbursable approach.

Larger group of bidders. Because it requires relatively little effort to prepare a reimbursable bid, the client can expect that all the desired bidders who have available capacity will submit bids.

Can be used in any market condition. When contractors are busy, they generally are willing to quote on reimbursable contracts. When they are slow, they are still glad to quote reimbursable rates, and the client can expect those rates to be substantially reduced. When a contractor's workload drops significantly, cashflow quickly becomes his first priority, and he will often be willing to bid for work at cost or even below cost in order to avoid losing personnel or suffer other unpleasant financial consequences.

Allows the scope of work to vary. Since the contractor will be paid for all work done, his scope of work can vary without causing contractual problems. This feature is well suited to the types of small projects whose scope of work simply cannot be well defined at the time the contract is negotiated.

Easier to negotiate. Less time and legal assistance is required for negotiation, and standard forms can be used. Many companies negotiate "umbrella" or "general service" agreements under which many small projects can be done.

There are also disadvantages to reimbursable contracts, as seen below.

Disadvantages of reimbursable contracts   The disadvantages of reimbursable contracts stem from the fact that the contractor has every incentive to expend the maximum number of man-hours and incur maximum costs, as every work-hour and purchase on the client's behalf carries with it some extra profit. Unlike the lump-sum contract in which the contractor takes a significant risk, the reimbursable contract puts virtually all the risk on the client. This means that the client has to do an effective job of project control if the desired results are to be achieved. And, in the small-project environment, the manpower, systems, and methods required for effective control may not be easily available.

The Right Approach

What, then, is the right contracting approach for a small project? For those small projects for which the scope of work can be well defined, the schedule will permit the time required for contracting, and during which market conditions are expected to be favorable, then the lump-sum contract should be considered. For those with a tight schedule and/or a variable scope of work, a reimbursable approach is probably appropriate. The necessary project-control techniques are described in subsequent chapters.

For many small projects, a contracting approach that is a combination of the best features of the lump sum and reimbursable contract may be appropriate, as described below.

Reimbursable Contracts With Fixed Fee

One way to reduce the contractor's incentive to expend the maximum number of work-hours is to fix the fee. In most contracts of this type, the fixed fee is based on the agreed-upon estimate of work-hours required to do the initial scope of work. As changes are issued and the scope of work increases, the fee is adjusted according to a negotiated formula, usually based on the estimate of work-hours associated with the change or increase in scope. This type of contract works well in most situations where a reimbursable approach is required.

For example, suppose it is agreed that the initial scope of work, as initially defined, will take 10,000 work-hours. If the all-in rate per hour is $50, of which the contractor's fee is $5, the fee for the initial scope of work will be $50,000, *regardless of the number of work-hours actually spent*. If the work-hours exceed 10,000, additional work-hours are reimbursed at a rate of $45 per hour; that is, without any fee payment. If additional work is authorized, the work-hours and fee are agreed beforehand.

Unit-Price Contracts

In unit-price contracts, a pricing schedule for performing specific physical tasks is negotiated. For example, if the work involves the welding of pipe the unit-price schedule would contain prices for welding of pipe of various diameters, flange ratings, and materials. As the work is defined and drawings are released for construction, each drawing is assigned a price based on the quantities shown, the number of tasks required, and the price per task. In this example, an isometric drawing would be priced according to the number of welds required in each category, and the agreed-upon unit-price sheet.

Unit prices can be thought of as lists of many lump-sum prices for very small packages of work. As such, it is similar to a lump-sum contract in which the scope of work can vary, and is therefore a useful form of contract for the small project. A big advantage is that contract administration and project control focus on physical quantities, not work-hours. One disadvantage of the unit-price contract is the difficulty of comparing bids and selecting the low bidder. Since the pricing schedule comprises many pages of detailed information, the mix (or "balance") of operations that is expected on a specific job will be a key factor in determining the low bidder. The only effective way to determine the low bidder is to make up a sample scope of work and price it according to each bidder's pricing schedule. Each bidder will allocate profit differently to the various operations being priced.

Most unit-price contracts, like most lump-sum contracts, also provide a schedule of hourly rates for use when the work scope is not covered by the unit-price schedule. When work is progressing simultaneously on both unit-price (or lump-sum) work and reimbursable work, the contractor will have to be watched closely to assure that hours covered by the unit price or lump sum are not also being treated as reimbursable.

Negotiated Contracts

When it is determined that there is only one contractor who can perform the required work, the contract—which is then negotiated on a noncompetitive basis—is a "negotiated contract." Many small projects are done this way, due to the use of preferred contractors, or those with specialized skills. Despite this, the client need not be in a weak position.

The first rule of negotiated contracts is to avoid jumping to the conclusion that there is only one acceptable bidder. Although it may mean extra work to identify and qualify one or more new bidders, the results may well be worthwhile. Project leaders often go to a single contractor simply because that contractor has done similar work before and is known to be reliable. Although those are good reasons for selecting a contractor, the company's best interests are served when several qualified contractors are available for anything that needs to be done. And, the best way to qualify a contractor is to have him do some work for you, so the assumption that a single-source contract is necessary can be challenged.

In those cases where a negotiated contract is necessary, the same principles should be applied as those used in a competitive situation. The client can estimate what the contract pricing schedule would be if a competitive situation existed by using past contracts and current-cost data as a guide. The client can also ask the contractor to provide "open book" details of his bid, to show how the pricing was arrived at, and why it is fair. Finally, he can use the strong bargaining position he has as the client, whose continued business is valuable to the contractor.

Incentive Plans

Another means of creating the incentives to perform that are associated with lump-sum contracts is through the use of incentive plans. Although these are generally associated with larger projects having complex contracts, there are many small-project situations where an incentive plan is useful. For example, a critical maintenance or a major turnaround project, where a great deal of valuable production is lost for every hour the plant remains shut down, can benefit from an incentive plan in which the contractor earns a significant bonus for finishing on or ahead of schedule. Experienced project leaders know that the point of view of the client and contractor are quite different: each makes a profit in a different way, and each therefore views the project from his own profit-making perspective. The cost overrun that could seriously diminish the client's profit could be a boon to the contractor with a reimbursable contract.

The purpose of an incentive plan is to align the objectives of client and contractor by giving the contractor a profit incentive to do what also benefits the client. A good incentive plan should be designed so that it is quite possible for the contractor to earn a bonus. The benefits to the client should be such that he wants to pay the bonus as the cost of the bonus is far less than the financial benefit he derives from improved contractor performance.

Most incentive plans involve schedule performance, although costs, safety, and other project parameters can also be included. Incentive plans can be incorporated in any type of contract. There are two general categories of incentive plans.

Unilateral Incentive Plans

Unilateral incentive plans are not negotiated: the client simply makes the contractor aware that he will receive a certain bonus if he meets certain targets. For example, he might offer a bonus for completion on-schedule, with an increased bonus for every day that completion is ahead of schedule (see Figure 3.17). The unilateral incentive plan is relatively simple to administer because there is no negotiation involved: that is, it is a "take it or leave it" offer. As a result, such incentive plans usually provide a bonus for good performance, but no penalties if performance is not up to expectations.

Bilateral Incentive Plans

These are negotiated incentive plans, in which the client and contractor agree on every aspect of the plan and administer it in close cooperation. This type of plan can involve penalties for poor performance as well as benefits for good performance. It can benefit the client, providing greater incentive because of the contractor's involvement in its design and administration.

Cautions on Incentive Plans

Although a well designed incentive plan can produce an outstanding return on the effort and cost involved, incentive plans can also be counterproductive. One problem can be the contractor's preoccupation with the bonus to the point at which it becomes a distraction and too little attention is paid to more important matters. The incentive plan must be carefully designed and administered such that it is difficult but achievable. The benefit is clearly lost if it is too easy, and, if the target is too difficult, a lot of money and effort is wasted in pursuit of an impossible goal.

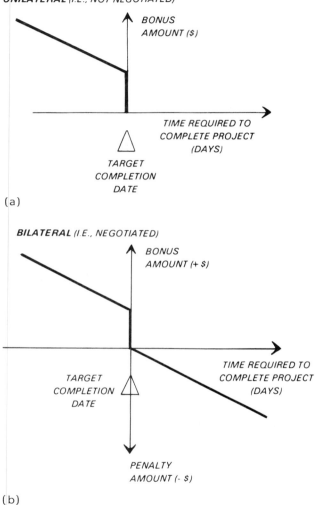

UNILATERAL (I.E., NOT NEGOTIATED)

BONUS
AMOUNT ($)

TIME REQUIRED TO
COMPLETE PROJECT
(DAYS)

TARGET
COMPLETION
DATE

(a)

BILATERAL (I.E., NEGOTIATED)

BONUS
AMOUNT (+ $)

TIME REQUIRED TO
COMPLETE PROJECT
(DAYS)

TARGET
COMPLETION
DATE

PENALTY
AMOUNT (- $)

(b)

Figure 3.17 Incentive plans. (a) Unilateral (i.e., not negotiated). In this example, the contractor will get a bonus for meeting the target completion date. For every day ahead of the target completion, he receives an additional bonus. There is no penalty for failing to meet the target date. (b) Bilateral (i.e., negotiated).

Steps Involved in Contracting

The project plan should allow ample time and resources to perform the work required for contracting. A properly executed contract plan will contain the following steps:

1. *Screening of contractors.* This is the identification of those contractors who are both able and willing to bid and perform. This is particularly important when a lump-sum bid is planned. The screening can be done by letter, telex, or even by telephone, using a standard questionnaire.

2. *Selection of bidders.* Based on the list of potential bidders generated in step 1, the final list of bidders is prepared. Factors in the decision as to which contractors will be asked to bid are:

Performance on past projects
Capability
Current workload
Interest in doing the work

When it comes to bid lists, many project leaders operate on the basis of "the more the merrier." In fact, there is little to be gained by having too many bidders, except a lot of extra work and lost time. While there is no rule as to how many bidders are appropriate in a given situation, the project leader should be guided by the principle that a Request for Proposal should be given only to those contractors who are best qualified to do the work.

3. *Preparation of the Request for Proposal.* The request for proposal should be carefully prepared and state, as specifically as possible, what will be required of the contractor both in his bid and, should he get the contract, in his performance of the work. The more the contractor knows of what is expected, when bidding, the less excuse he has later for not performing. The Request should in all cases contain:

A description of the scope of work (general for reimbursable contracts, specific for lump sum)
The reporting requirements for project control (discussed in subsequent chapters)
The specific format for the bid pricing (to assure that bids can be compared on an consistent basis)

4. *Bid analysis.* This involves the reviewing of the bids to select the preferred contractor. Bid review is enhanced considerably when a check estimate has been prepared. Each bid can then be compared in every re-

spect to the check estimate, and errors, omissions, or things that seem out of line can be identified and, if necessary, corrected. The bid review really consists of two separate analyses:

"Hard-money analysis": This is the quantitative analysis of pricing, compliance with contract items, and other commercial aspects.

"Soft-money analysis": This is the qualitative analysis of contractor capabilities, based on past performance, quoted performance, and specific evaluations by the client. Using a rating sheet the relative technical and management capabilities can be ranked.

In some cases, the choice turns out to be difficult, such as a contractor whose performance is predicted to be above expectation, but whose price is not the lowest. Anyone who has ever tried to save money by buying the cheapest product knows that the low bidder is not necesarily the best choice, but this does not make it easy to explain to management why the low bidder was not chosen. One way to quantify the judgement process in such cases is to use the project model. For example, the network durations can be changed to reflect the anticipated schedule performance of each contractor, and the net effect on schedule of selecting each contractor can be evaluated. Then, the economic benefits of the improved schedule performance can be compared with the increased contract costs. Or the economic model used to justify the project can be run to simulate the results of using different contractors with varying design capabilities, by varying operating costs and service factors.

5. *Contract award.* This final step involves the negotiations with the selected contractor. This is the last chance for the client to negotiate from a position of strength; once the contract is signed the contractor, as discussed earlier, has the stronger position. Many clients overlook the importance, at this point, of defining the reporting requirements that will assure project control: such things as progress measurement, manpower, reporting cycles, etc. *The essence of effective contractor control is the contractual obligation to provide information that can be used by the client for unbiased performance measurement* (see Chapters 11 and 12). Such information is seldom offered voluntarily.

PLANNING FOR CONTROL OF TECHNICAL QUALITY

Project control, as pointed out in Chapter 2, is really a process of optimization of the cost, schedule, and quality to achieve the maximum profitability. Although most of this book deals with the control of cost and schedule, quality control is an equally important aspect which affects them both (see Chapter 10 for further discussion).

What is "quality"? From a project engineer's standpoint, the relevant terms are defined as follows:

Quality: *conformance to requirements.* Using this definition, it can be seen that the maximum quality for an engineered item is not necessarily the use of the most expensive materials, or the latest technology, or hand-fabrication. The only attributes that add to quality are those that make it more fit for its intended purpose, which is usually to enhance profitability. This simple definition, when applied to proposed design changes, can help separate those that are necessary from those that are not.

Quality assurance: *the plans, standards, and methods by which the work will be controlled to assure satisfactory quality.* On most projects, quality assurance requires a definition of quality in terms of specific standards (such as building codes, API specs, etc.) that will be enforced. A recent trend in defining quality in specific design terms is to consider it in terms of life-cycle costs (see Chapter 10). Since some performance standard is expected of every part of the facility over its entire operating life, it is reasonable to look at design vs. cost evaluations not only in terms of investment cost, but also in terms of the impact that a design change could have on operating costs such as power consumption and maintenance. The first step in quality assurance is, therefore, to define design quality in specific performance terms, as well as in terms of design criteria.

Quality control: *the actions that assure acceptable quality.* Typical quality-control activities include inspection and testing of manufactured parts and assemblies, welding inspection, certification (when appropriate), and design reviews. Design reviews are a recent trend that has been shown to aid significantly in reducing late changes and assuring fitness for the intended purpose. Design reviews require that the ultimate users of the facilities—usually operations and maintenance staff—participate in periodic reviews of the design and construction, be made aware of current design problems and decisions, and make any input to the design process that they feel appropriate. If certifying bodies are involved, their reviews must also be scheduled.

If effective quality control is to take place, the necessary activities must be shown on the network plan and schedule. This is another project-management function that is often overlooked in the planning phase, causing problems such as late changes and rework. Typical quality-control activities that should be included in the plan are:

Design reviews (e.g., hazop, constructability, environmental)
Reviews by certifying authorities (if applicable)
Visits to fabricators and/or witnessing of tests

CHAPTER SUMMARY

This chapter described the techniques available for planning and scheduling. Important considerations that affect the plan, such as contracting strategy and quality assurance, were also described.

The network plan and barchart schedule that have been developed for small projects, using the methods described so far, is the foundation of all that follows. Scheduling the use of resources, cost estimating, requesting project approval, obtaining commitments of resources, and implementing project control will all be based on this project plan. Although our plan and schedule may change as the project progresses, this original plan will always be our basis for comparison. We will always measure where we are by our distance from it. This original plan is our "static model" of the project.

4

Providing the Resources to Do the Work

WHAT IS RESOURCE SCHEDULING?

Resource scheduling is based on the inarguable premise that work cannot be accomplished without four essential resources: materials, people, equipment, and time. The project plan and schedule developed in Chapter 3 established the availability of time. If the project plan and schedule are to be achieved, it is now necessary to assure that the required material, labor, and equipment will be available when needed. The process by which this is accomplished is *resource scheduling.*

Although the importance of resource scheduling is self-evident, many projects suffer avoidable delays from inadequate resource management. This is especially true for multiple small projects, where many independent projects are competing for the services of a fixed pool of resources. So one of the most important benefits of our project model is that it makes it possible to do a credible job of resource management. Resource scheduling is the process of identifying the quantities of resources required each day and scheduling activities and resources so that the resources required never exceed the resources available. Resource scheduling consists of the following steps, each described later in the chapter.

1. Define the resources required to accomplish each activity on the network plan. We will refer to this process as *resource allocation.*
2. Define the total amount of resources required for the project, in each time period, according to the plan and schedule. We will refer to this process as *resource aggregation.*
3. Define the resources which can be made available during each time period, i.e., *resource availability.*
4. Compare resource requirements with resource availability to identify shortfalls, and reschedule activities to assure that resource requirements will not exceed availability. This is *resource leveling.*
5. Prepare a schedule for the provision of project resources. This *resource schedule* is usually presented as a *manpower histogram.*

Resource Planning Is Essential
for Small Projects

Although resource planning may seem an elaborate process to perform on a small project, it is a vital task. Resource planning is especially important for small projects, for the following reasons:

Multiple small projects are likely to experience resource shortfalls. Due to sharing a fixed pool of resources, each project must depend on the timely completion of other projects and the subsequent release of their resources. Since the scope and duration of small projects often increases, this causes real problems in coordinating and assigning resources between projects.

Small projects are profoundly affected by resource shortfalls due to the lack of alternative ways to make progress. On a large project, if a certain material is not delivered on schedule, there are usually many other work areas for which material is available and in which physical progress can be made. Similarly, the lack of a special piece of construction equipment, or of certain labor crafts, can be compensated for by working on other things. Since the large project has many noncritical activities, these adjustments to the plan can often be made with no net effect on the completion date. These alternative ways to make progress often do not exist for small projects. A resource schedule helps to avoid shortfalls, but, when shortfalls are experienced, the resource schedule makes it easier to identify alternative courses of action.

Small projects can often suffer delays due to resources not being available when needed simply because they have not been identified. A resource schedule assures that all necessary resources are identified well in advance and commitments made to provide them on schedule.

Small projects are often likely to experience resource shortfalls due to changing priorities. In an operating plant, where project work may not be a top priority, resources such as plant labor, equipment, or warehouse supplies may be diverted to higher priority uses just when they are needed on the project.

Project leaders often find it difficult to obtain the necessary commitments from other parts of the organization to provide resources. The resource schedule represents a clear commitment that is agreed upon before the project begins, and which can be compared with actual performance. Then, when promised resources are denied, the project leader can show the effect that the resource shortfalls have on the schedule; this can help prevent the shortfalls from occurring in the first place.

DETAILS OF RESOURCE PLANNING

Step 1: Resource Allocation

Each activity in the plan is reviewed to determine the resources required in each category to accomplish the specified scope of work in the time provided by the activity duration. The easiest way to accomplish this is by using our PC-based project management software to input resource requirements as we define each activity. It should be noted that the duration is itself a direct function of the amount of resources applied. When computerized project models have been used for some time, disk files of actual project experience can be used as a guide to determine the time and resources required to perform typical tasks.

For small projects, it is generally assumed that the number of resources in each category is constant over the duration of the activity. Where this is not the case, a resource profile for each activity can be established. This type of analysis poses no special problem, but it requires a more sophisticated computer system than is apt to be available for the small project. Proceeding activity-by-activity, the network-resourcing is completed. It might seem that, if we were to stop the resource analysis at this point, we would have a reasonable resource schedule for the project. However, the next step often reveals that, when total project resources are considered, there are sharp peaks in the resource requirements that exceed the availability.

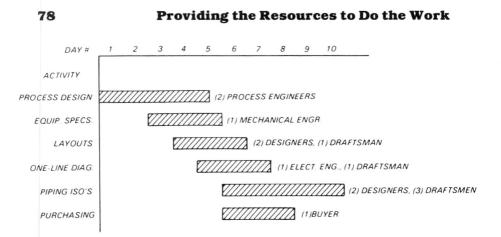

Figure 4.1   Resource allocation. In this example, the bar chart shows the design of a small process-plant project.

Step 2: Resource Aggregation

Given the number of resources in each category required to complete each activity, and, given the scheduled time period for each activity, it is a simple matter for a computer to calculate the total resource requirements for all activities occurring during each time period. In this manner a histogram of total project manpower can be obtained, as well as one for total manpower in each discipline (see Figure 4.1 and 4.2). Resource aggregation would be a chore if done manually, but our project management software can do this instantly. The result of resource aggregation,

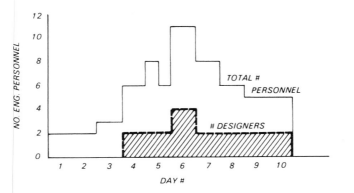

Figure 4.2   Resource aggregation histograms derived from Figure 4.1.

showing total manpower requirements per time period, is a very useful tool:

1. The manpower histograms are essential in determining each project's requirements and in scheduling it within the population of other small projects. We can see that the project shown in Figure 4.1, which requires two process engineers to start, can begin as soon as preceding projects release them. Note that a resource histogram provides vital information, yet is not intuitively obvious. It is therefore one of the most useful results of the computerized project modeling calculation.

2. The total manpower histogram can be converted to a cumulative manpower curve, or "S curve," as shown in Figure 4.3. The S curve for manpower is useful in project control as a means of comparing the actual expenditure of manpower against the original plan.

3. The cumulative manpower curve can also be used to derive the S curve for physical progress. This curve, showing cumulative progress as a

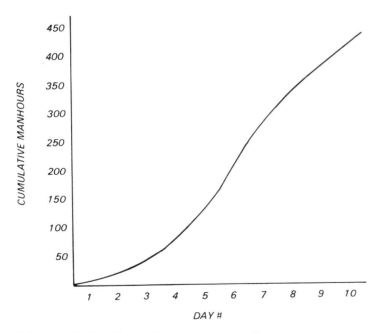

Figure 4.3 Cumulative manpower vs. time.

function of time, is probably the most important project-control device, and is an important part of the plan. If we make the reasonable assumption that productivity is constant over the life of the small project, the progress curve will have the same shape as the manpower curve, and is derived simply by assuming that the percentage of planned direct work-hours expended at a certain point in time is equal to the percentage of planned progress. (Note that we are dealing here only with planned, not with actual, work-hours and progress. This subject is explored in much greater depth in Chapter 12.)

4. The cumulative manpower also provides a good basis for calculating expenditures, since most of the cost of a small project is likely to be labor-related. Expenditure forecasting is a particularly time-consuming task that is important but contributes little to project control, and any short-cut approaches are usually very useful.

5. The cumulative manpower and progress curves are also useful for small projects because they often do not resemble the traditional "S" shape that is generally seen on large projects. The S shape results from the fact that most large projects begin with activities, such as civil work, which must precede the rest of the job and yet can only utilize a certain number of people. Once the sitework is done, the underground piping and ducts installed, and the foundations complete, then many other activities can commence. Similar phenomena also exist toward the end of the large project.

Since many small projects, such as turnarounds, maintenance, and other in-plant projects, start with a full workforce and work in many areas at once during the project's short duration, the shape of the cumulative manpower and progress curves is impossible to predict without some sort of resource aggregation procedure.

Step 3: Defining Resource Availability

Manpower Availability

In the multiple small project environment the availability of resources can be limited by a number of factors including:

Organization/budget limitations: Small projects that must draw manpower resources from a labor pool that is shared with other projects (e.g., drawing office, plant mechanical forces, or fixed company staff), and that must rely on manpower to be released from preceeding projects, will be likely to experience limitations in manpower availability due to the need to share resources with other projects.

Manpower density: Many types of work are limited by the number of
workers that can work efficiently and safely in a given area. When
manpower densities exceed the guidelines, productivity is likely to
suffer due to interference, supervision, and logistical problems. Safety
can also be a reason to set limits on manpower density.

Physical constraints: Some operations can absorb only a certain number
of workers due to physical limitations. For example, the erection of
the frames for a heavy steel structure can utilize only a limited number of steel-workers.

Limited availability of special skills: Some projects require specialized
skills that are in short supply, either in-house or from contractors
or consultants. This problem is particularly acute for projects in remote locations, where the local population cannot supply the needed
skills, and the number that can be transported in is limited.

Logistics due to project location: Some projects, best exemplified by
offshore oil platforms, have severe limitations placed on the availability of manpower simply by the logistics involved. Those who
have worked offshore know that the number one factor in determining manpower availability is the number of beds available on the
platform. Other logistical factors such as supply boats and helicopters
play a big part in determining the manpower available.

Material and Equipment Availability

Limitations on availability also exist for material and equipment. Material resources on small projects can be limited in availability due to delivery schedules, warehouse stocks, or project priority. Small projects
often suffer from material shortages caused by preceeding projects taking materials intended for the project at hand. Remote projects, of course,
have material availability problems stemming from the logistics involved
in receiving, storing, and transporting materials to remote sites. In some
cases, such as offshore projects, storage on-site is limited as well.

Equipment availability can be limited by the requirement that in-
plant equipment be used, and the necessity of sharing the pool of available equipment with other projects. Often the failure to properly define
project priorities can lead to heated debates about where a particular
piece of equipment ought to be used next. Equipment availability can
also be limited by the physical constraints of the site, as well as by the
need to avoid congestion and interference with the operating plant.

After careful consideration of the above factors, and discussions
with those who will be responsible for providing the resources, the project

leader is able to input resource availability for each type of resource category as shown in Figure 4.4. By overlaying the resource-availability histogram on the resource-requirement histogram, the resource shortfalls in each category can be seen.

For the project to be completed according to plan, it is now necessary to adjust the schedule to bring resource requirements into line with availability.

Step 4: Resource Leveling

The process of resource leveling, usually done by our project management program, proceeds as follows:

The *early-start schedule,* in which each activity is assumed to start on its early-start date, is *used for the initial resource aggregation.* Thus the initial histogram can be thought of as an "early-start" histogram. By comparing requirements with availability, resource shortfalls are identified.

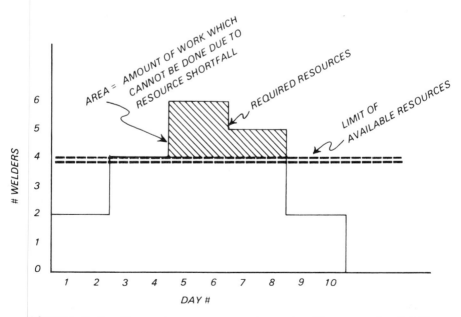

Figure 4.4 Comparing resource requirements with resource availability.

The first attempt to eliminate resource shortfalls is made by delaying the start of various activities. This can be thought of as allowing non-critical activities to move within the time envelope of their duration plus float (see Figure 4.5).

If resource shortfalls still exist, it is necessary to allow either the schedule for the project to slip or to increase the available resource levels (see Figure 4.6). If the project-completion date cannot be allowed to slip, we have a case of *time-limited* resource leveling, in which we simply identify the time period and amounts that various resource availabilities must be increased. If the resource availability cannot be increased, we have a case of *resource-limited* leveling in which we must allow the schedule to slip in order to accomodate the available resources.

From the above it can be seen that resource leveling, like scheduling, is very much an iterative process in which the goal is to find the optimum combination of the project parameters of time and resources.

Resource Leveling by Interaction with the Project Model

The resource-leveled schedule calculated by project management software, as described above, may not represent an optimum solution. Activities may be delayed more than they need to be, due to the software's inability to make human judgements. So, while it is true that a leveled schedule

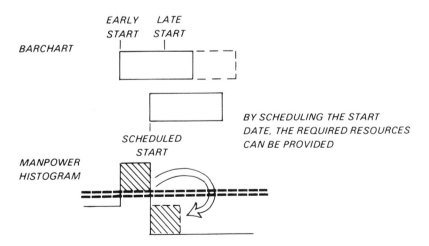

Figure 4.5   Using float to eliminate resource shortfalls.

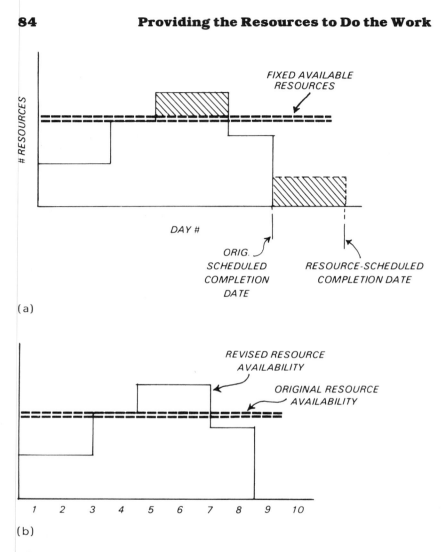

(a)

(b)

Figure 4.6 (a) Resource- and (b) time-limited scheduling. Resource-limited scheduling allows the schedule to slip to match availability while time-limited scheduling allows resource availability to match requirements.

Figure 4.7 Schedule of resource requirements. (a) Project engineers, (b) Designers, (c) Laborers, (d) Welders, (e) Mobile cranes, (f) Welding machines.

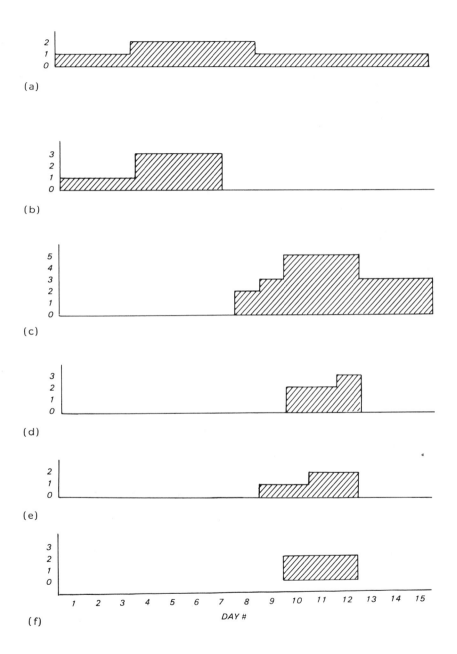

(a)

(b)

(c)

(d)

(e)

(f)

will not have any periods in which resources are overloaded, it may also be true that there are periods in which resources are underutilized.

It is therefore important to interact with the project model in order to find a better compromise between limits on resource availability and the need to get things done. Some of the ways that the plan can be edited to improve the schedule of activities and resources are:

extending duration to reduce resource requirements

splitting up activities that cause resource overloads

changing the dependencies (using soft logic, overlapping activities or both) to move some work into periods of spare capacity

increasing resource availability by changing the project or resource calendars—to add overtime work, shifts, weekends or holiday work.

temporarily increasing availability by the provision of extra people, subcontracting some work, hiring temporary help, transferring people from other projects, etc.

substituting a different (underutilized) resource for an overutilized one.

When we interact with project management software, leaving resource leveling on, the program will recalculate the resource-leveled schedule each time, thereby showing us the improvements made by our changes (see Chapters 7, 15).

The result of these calculations is a resource-leveled schedule, the only schedule in which we can have confidence.

Step 5: Schedule Resource Requirements

With the resource leveling complete and the resourced schedule finalized, it is appropriate to prepare a schedule of requirements for those who will be responsible for providing the necessary resources. These resource schedules can take the form of a histogram or even a simple list showing what is required and when. The resource schedule should then be used to obtain agreement and a commitment to provide the resources as scheduled. Examples are shown in Figure 4.7.

ALLOCATING RESOURCES TO MULTIPLE PROJECTS

One of the biggest problems of small projects is the management of many projects at once. Perhaps the greatest difficulty involved in managing multiple projects is that of deciding how to allocate resources among them.

For example, the manager of the project-engineering department may have a "pool" of 10 project engineers, each of whom is handling 10 projects. A new project comes into the department, and she must decide to whom this project should be assigned. Or, the drawing-office manager may have just so many designers and draftsmen; how should they be allocated to the various projects? The program manager or project leader has a similar problem. These resource pools usually also exist for construction labor and equipment, as well as for maintenance.

The solution to this problem lies in the concept of network hierarchies, as discussed in Chapter 3, combined with the principles of resource aggregation. After resource planning on an individual basis at Level 2, the project can be represented as an activity on a simple Level 1 network, whose resources are summarized. The several projects under consideration can then be handled as if they were diverse parts of a single large project, which, in fact, they are. The same principles of resource allocation can then be applied to the family of projects at Level 1.

In the example shown in Figure 4.8, we have 10 engineering projects, each represented by one activity on the master network. This master network represents a "project" that consists of all the current and planned small projects we are managing. For example, it might represent the current plan for all the projects to be done this year. Although many projects are independent, there will be some cases in which one project must be complete before the next one can start, and the network reflects those dependencies. What, then, will determine when each project gets done? There are usually three factors:

The project's priority
The constraints shown in the network
Resource availability

While the priorities and constraints may be out of the project leader's control, he or she can affect how the resources are deployed. If a normal resource analysis is performed on this multi-project network, resource aggregation will indicate which projects are likely to be delayed because of shortfalls, and resource leveling will indicate how the projects can be scheduled to match the available resources in the pool. When new projects are added and completed, and when priorities or the availability of resources change, the analysis can simply be updated. Such an analysis can be useful in justifying the need for outside services, such as subcontracted engineering work. Many project-management software packages offer this multiproject capability (see Chapters, 7, 15).

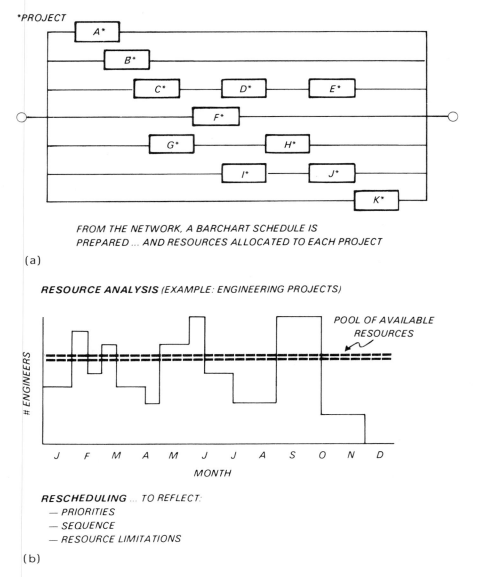

Figure 4.8 Allocating a fixed pool of resources to multiple projects. (a) The network shows each project represented by one or more activities. A, B, C,...,K represent these projects. From the network, a bar-chart schedule is prepared and resources allocated to each project. (b) Resource analysis (in this example: engineering projects). Rescheduling reflects priorities, sequence, and resource limitations.

This technique is particularly useful for planning and decision making at the management level. It can be used to address problems caused by several projects competing for limited resources, problems that arise when the available resources for a project have to be increased, and also to help set priorities.

CHAPTER SUMMARY

In this chapter we discussed the importance of resource planning to multiple small projects, and showed how it can be done. The original schedule may have changed as a result, but we can now be confident that we know what resources have to be provided so that the plan can be executed according to schedule. And, we can be confident that our resource-leveled schedule is realistic. Our plan, schedule, and resource analysis have provided us with a model of the project that we can use for estimating, decision making, communication, and subsequent project control. And our multiple-project plan, activity schedule and resource schedule helps a multiple-project leader or resource manager keep deadlines in line with priorities and staff limitations.

5

Cost Estimating for Small Projects

IMPORTANCE OF THE COST ESTIMATE

The cost estimate for a project serves two important functions:

It is the basis on which the *economics* of the project are calculated, and therefore an essential element in the decision to go ahead.
It serves as the basis for *cost control*.

In spite of the importance of the cost estimate, most small projects suffer from a lack of accurate, consistent, well-organized estimating. As mentioned in Chapter 1, this is often due to the inherent difficulty in estimating small projects involving revamp work, as well as to the lack of trained estimators and formal methods and data. Unfortunately, small projects have a tendency to overrun due to variations in the scope of work, unforeseen difficulties, etc., and, when they do, the overerun percentage is apt to be great because the original estimate is small. To put it another way, management often fails to realize that it doesn't take much to cause an overrun of greater than 10% on a small project. And in spite of good work by everyone, a project that finishes more than 10% overbudget is often regarded as a project management failure.

It is, therefore, evident that a better way to estimate small projects is needed: one that provides greater consistency and simplicity; provides a framework for capturing actual cost data for use in subsequent estimates; addresses the difficulties of estimating labor on revamps; and provides a good basis for cost control. This chapter describes the concepts and methodology necessary to accomplish those goals.

THE INTEGRATED APPROACH TO ESTIMATING

Chapters 3 and 4 describe the methods by which we create a model for the project that defines:

The *activities* that must be performed in order to complete the project
The *sequence* in which the activities will be performed
The *time period* in which each activity will be performed
The *resources* that will be applied to each activity

After the planning, scheduling, and resource leveling have been completed, all the project parameters are in balance, that is, the resources that will be applied are adequate to accomplish the defined scope of work of each activity in the scheduled time. It now remains to estimate what it will cost to provide those resources. It is therefore appropriate to use the resource plan as the basis for estimating labor and equipment costs.

The basis of the estimate in the integrated approach to cost estimating derives from the project model. For each activity, the cost is estimated:

To do the *work* required
In the allowed *time*
With the allocated *resources*

This technique is particularly appropriate for small projects as the labor-related costs represent a major and often ill-defined part of the project cost. The estimating process is also greatly simplified, and is well suited to situations in which a lack of relevant historical data is available and judgement must therefore be used.

The integrated approach will be described in detail in this chapter. However, the basic principles of cost estimating also apply to small projects, and these are discussed below.

PRINCIPLES OF COST ESTIMATING FOR SMALL PROJECTS

What Is a Cost Estimate?

Before exploring further the concepts and techniques of cost estimating, it is appropriate to define exactly what we mean by a cost estimate in the context of a project. A cost estimate is:

1. "A calculation of the *approximate* cost." This definition, from Webster's dictionary, reminds us that, contrary to what is popularly believed by those who approve budgets, an estimate is an approximation. When functioning as estimators we need to remember this and make sure those who work with our estimates remember it too. There are ways to measure the accuracy and uncertainty in an estimate, and these will be discussed later in the chapter, as well as in Chapter 6.

2. A *forecast* of the final cost for the project, under an assumed set of conditions. This part of the definition reminds us that an estimate is based on assumptions. Even if we have some "firm data" on design or cost, we are assuming that firm data will remain firm for the life of the project. For our estimate to be exactly correct, every assumption, whether explicit or implicit, must prove to be correct, and this very seldom happens. So, we should look for an estimate to be approximately correct, not exactly correct. All the basic assumptions should be reasonably close to correct, and those areas that were underestimated should be offset by those that were overestimated.

This definition also reminds us that an estimate is a *prediction*, or forecast, of what will happen over the project's life. The extent to which information is undefined at the time of the estimate is predicted by the estimate according to the assumptions made and the methods used. Estimate accuracy increases as the amount of definition increases, and the amount of prediction decreases (see Figure 5.1).

3. A *reflection of the plan* for the project's design and implementation. The cost estimate for the project can be thought of as a "budget" in the literal sense of the world. In financial planning, the word budget implies "plan"; that is, the budget is the plan for how money will be spent. Our project-cost estimate is also such a plan, and, as such, must reflect to the greatest extent possible how the work will be done.

Characteristics of a Cost Estimate

Unbiased

Unless stated otherwise, the estimate should have an equal probability of overrun or underrun. This requirement is especially important for multiple small projects. The reason for this is simple: if a company is funding 100 projects per year, and, if each is estimated at the 50/50 probability point, the odds are that the money set aside for all the projects will be about right. If, however, all the projects have conservative estimates, with extra contingency, the odds are that too much money will end up being set aside for the projects. As a result, the unspent project funds will

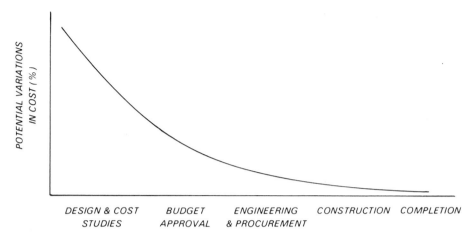

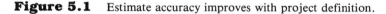

DESIGN & COST   BUDGET   ENGINEERING   CONSTRUCTION   COMPLETION
    STUDIES   APPROVAL   & PROCUREMENT

STAGES OF PROJECT DEFINITION

Figure 5.1   Estimate accuracy improves with project definition.

not have been available for other investments or expenditures. Anyone who has completed an internal project at a large cost underrun and expected thunderous acclaim has probably found that the reaction to having the funds returned was not so good, since it was correctly perceived that those funds could have been put to good use elsewhere.

In the case of estimates for bidding, or for particularly cost-sensitive projects, the 50/50 estimate may not be the most desirable. In the case of a bid, the probability of getting the work must be weighed against the probability of overrunning the estimate. In the case of a cost-sensitive project, the project may be necessary but risky so it is therefore desirable to set aside sufficient funds so that there is a low probability of overrunning. In either case, however, it is important to assure that any bias away from the 50/50 probability point is clearly documented and understood.

Appropriate to the Project Definition

Many companies make the costly mistake of creating an estimate at a level of detail that is far greater than the level of reliable design information available. This practice produces an estimate of greater cost but the accuracy is not improved. It also fosters the dangerous illusion that the estimate is more accurate than it really is. *The fact is, an estimate can be no better than the information on which it is based.* If a project is in the early design stage, the preliminary nature of the design basis makes any estimate prepared at that stage a preliminary estimate, no matter how much detail may be included. Therefore, the method used, the time and cost, and the basis of an estimate should all be appropriate to the project's status and to the purpose for which the estimate is intended. This is illustrated as follows:

Early-planning estimate   This kind of estimate might be used to make an initial evaluation of a project, or to rank potential projects and select those for further work. Relative accuracy is more important here than the absolute value of the estimate, so quick, approximate estimates are appropriate. Other techniques could include estimates made by adjustments to data from previous similar projects, as well as those based on "rules of thumb," and use of the "Six-Tenths Rule."

Budget estimate   This kind of estimate is used to set the budget for the work, calculate profitability, obtain management approval, and acts as a basis for cost control. If detailed design information is available, the estimate should be done in the same level of detail. If not, the estimate should reflect the information that is available and considered firm.

Establishing the Estimate Basis

One of the most important concepts in cost estimating, especially where small projects are concerned, is establishing the basis of the estimate. A cost estimate can be thought of as resting on a "tripod" (see Figure 5.2a). Each "leg" of the tripod is described as follows:

The Design Basis: What Is to Be Built

The design basis usually takes the form of drawings, sketches, and specifications. The most significant things about the design basis are the level of definition and the potential for change. Many small projects, especially those involving modifications to existing facilities, are defined in great detail by the design group and appear to be on a firm basis. However, once the project progresses further and problems become apparent, major scope changes are likely to occur. For example, the original design often assumes that facilities can be installed in the most straightforward way. Sometimes the facts turn out otherwise. Therefore, the potential for change is as important as the level of detail of design definition.

Figure 5.2(b) illustrates that the amount of definition of the basis determines the estimate accuracy: the more that is defined, the less the estimate must predict, and the more "solid" the basis.

ALL VARIATIONS IN PROJECT COSTS CAN BE EXPLAINED
BY VARIATIONS IN THE DESIGN, PLANNING & COST BASIS

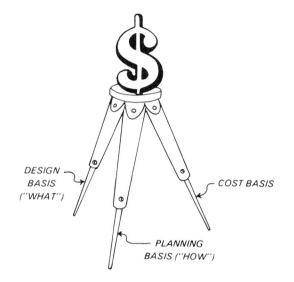

AN ESTIMATE WILL BE ACCURATE TO THE EXTENT THAT
IT CORRECTLY PREDICTS THE UNDEFINED PART OF THE BASIS

(a)

Figure 5.2   (a) The project cost stands on a "tripod" base. All variations in project cost can be explained by variations in the design, planning, and cost basis. (b, left) The early planning estimate must predict most of the basis; (b, right) The detailed estimate has most of its basis defined.

The estimate will be correct to the extent that it correctly reflects the design basis and predicts the extent of changes.

The Planning Basis: How It Is to Be Built

The planning basis usually consists of the network plan, barchart schedule, and resource plan as defined in Chapters 3 and 4. The relationship between how the project is managed and its cost has now been discussed sufficiently that it can be seen that the estimate should reflect the contract plan as well as any costs associated with maintaining or accelerating the schedule.

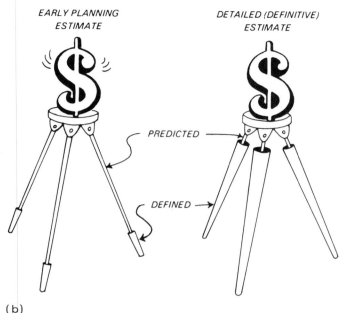

(b)

Figure 5.2 (Continued)

The estimate will be correct to the extent that is correctly reflects the planning basis and predicts how the project will actually be built.

The Cost Basis: What Quantities and Pricing Levels
Will Be Experienced

The cost basis is the cost data and estimating methods used to prepare the estimate. It consists of the data used for estimating, the assumptions made as to escalation, productivity, and other estimating variables, and the judgements made by the estimator. Since most estimates are done at some base date and then escalated, the basis should also define the base cost level.

The estimate will be correct to the extent that it correctly reflects cost data and predicts what quantities, pricing and performance will actually be experienced.

The estimate basis should be documented carefully and, if possible, be approved before the estimate is prepared. Assumptions, in particular, should be well documented, as well as calculations based on judgement. Documentation is important to enable reconciliation of cost estimates and forecasts with the original budget (see "Reconciling Forecasts and Estimates" and see estimate preparation).

The Estimate Is the Basis for Cost Control

Like the plan and schedule for the project, the estimate's most important function after project approval is to permit cost control. In order to be effective as a control device, the format of the estimate must be closely matched to the way the project will be run. For example, it should be easy to group costs into the major contracts and subcontracts, and to use the cost codes for aggregation and reporting of cost data from contractors and other sources. The integrated approach serves this purpose well. In cases in which the estimate used for project budgeting and approval is not broken down into sufficient detail for control, many companies find it is worthwhile to create a control estimate with the necessary detail.

Reconciling Forecasts and Estimates

A very useful tool in cost control that is frequently overlooked is the use of reconciliations. An estimate reconciliation is simply a comparison of a past estimate or forecast to a current one, and an explanation of the differences. Through this process, trends can be brought to light that might otherwise have been overlooked. In the multiple small-project environment, reconciliations can be a good shortcut method for keeping track of things and keeping the right people informed. Management, which must always be concious of the budget, always appreciates good reconciliations. Unfortunately, good reconciliations are seldom seen, because the principles of reconciliation are not widely understood. Useful reconciliations compare:

Current cost estimates against the previous estimate (this indicates major design changes)

Current cost forecasts against the previous forecast (this indicates recent trends)

Current cost forecasts against the current-control estimate (this indicates performance trends)

Current cost forecasts against the budget (this indicates changes and developments since project approval)

Of these, the two most useful compare the current forecasts or estimates against the previous forecast or estimate (i.e., how have things changed since we last looked at the cost?) and current forecasts against the budget.

A useful form of a reconciliation is shown in the following example:

Cost (1,000s)		
Previous estimate		100
Design variations		+ 14
Added piping	+ 6	
Revised electrical	+ 5	
Alloy materials	+ 4	
Net all other design variations	(1)	
Planning variations		(5)
Subcontracted drafting	(3)	
Overtime hours	+ 2	
Shared field facilities	(4)	
Estimating variations		(1)
Equipment pricing	+ 2	
Market conditions (piping)	(1)	
Steel quantities	+ 4	
Contingency reduction	(9)	
Net all other	+ 3	
Current cost forecast		108

It can be seen that this reconciliation format shows the variations to the project cost, broken down in terms of the three legs of the estimate basis: the design, the planning, and the cost basis. All variations to the estimate fall into one or more of these categories. A reconciliation often shows that, although the total cost has not changed significantly, the variations in the basis are quite significant and could be a sign of greater variations to come. In addition, the nature of reconciliations is such that they reveal problem areas that might not otherwise show up.

In preparing a reconciliation, the calculation of the cost impact due to each variation is done on the same basis as the original estimate and assumes that nothing else has changed. Thus, each line should show how the original estimate would have been different if the single variation at hand had occurred.

DEFINITION OF ESTIMATING TERMS

As in planning and scheduling, the proper definition of terms is most important in cost estimating. Unfortunately, the terms used in cost estimating are used so often in other contexts that they are often ambiguous. It is therefore essential that we begin our discussion of estimating techniques by defining the appropriate terms.

Project-cost estimate: This is the estimated cost to complete the scope of work defined in the project. This definition can be difficult when the question becomes, "Where does the project begin and end?" The cost estimate should clearly specify the scope of work by stating the design basis and by specifying inclusions and exclusions so that it is clearly understood what is and what is not covered by the estimate.

Capital costs: Many small projects are "capital projects," that is, they result in a facility that is treated as a "fixed" depreciable asset.

Expense costs: These costs are those that can be considered to be necessary for the operation of the existing facilities. The confusion that often exists over capital and expense costs comes from the fact that many small projects include both types of costs, as well as from the fact that the tax implications of these costs adds a level of complexity. For example, the costs to build a new facility are clearly capital costs, but the costs to start it up are often considered expenses. There is often a good deal of confusion on a given project as to how these costs should be classified, and the estimator needs to have these questions resolved if effective cost control is to take place.

Direct material costs: This is the cost of the material that is required for the fixed project facilities, is not consumed during the project, and is not reused elsewhere. Piping, vessels, buildings, and pumps are examples of direct materials. Catalyst, replacement parts, warehouse spares, scaffolding, concrete forms, and welding rod are examples of indirect materials.

These costs are generally divided into two categories:

1. *Equipment* refers to those items that produce some process change. For example, pumps, heat exchangers, pressure vessels, and compressors all have, as their purpose, the changing of pressures, temperatures, and other process variables. In a processing facility, such items as generators and transformers would not be considered equipment because they are part of a utility system. However, in a company that builds power systems, the transformers and generators would be considered equipment. Once again, the important thing is to define the terms for company use and then stick to a consistent definition.

2. *Bulk materials* (sometimes referred to as materials) covers those items that are necessary for the equipment to function. Examples of bulk materials are piping, electric cable, foundations, and structural steel.

Direct labor: Direct labor is the labor that produces measurable physical progress. This definition intentionally rests on the method used to measure progress. People like to compare actual work-hours spent with measured progress and draw conclusions about productivity, so the work-hours and progress measurement must be consistent. Direct field labor generally includes welders, electricians, laborers, etc. For engineering work, direct labor includes designers and draftsmen.

Indirect labor: Indirect labor is required to make it possible for direct labor to take place. Indirect labor in the field usually includes supervisors, timekeepers, warehouse attendants, etc. For engineering work, indirect labor includes management, project control personnel, and office staff. Indirect labor is associated with a specific project.

There are several labor categories whose definition as direct or indirect labor frequently causes confusion. These include:

Foreman
Material handling
Testing (e.g., x-ray)
Scaffolding
Purchasing

These, and any other areas of ambiguity, should be clearly defined before the work begins.

Overhead costs: Overhead costs are not project-specific, therefore they tend to be prorated over all multiple projects. Examples of overhead costs are computers, training, functional management, and office support staff.

Contingency: There is so much confusion and ambiguity regarding contingency that it has been given a separate section in this book (see "Estimating Contingency" later in this chapter, and Chapter 6). Suffice to say at this point that contingency may be defined as the provision for those variations from the estimate basis that are likely to occur but that cannot be specifically identified at the time the estimate is prepared.

Escalation: This is the change in price levels over time. For a project cost estimate, the calculation of escalation requires definition of a "base date" that reflects the time period at which the prevailing prices in the estimate existed. The estimator must also define the "centroids" of material, labor, engineering, and other cost categories that represent the average point in time at which procurement of those items will take place. Escalation "factors" can then be applied to bring the estimated costs up to the pricing level that is expected to prevail at the time of the project (see Figure 5.3 and "Estimating Escalation," later in this chapter). "General escalation" refers to the escalation that exists for the industry in gen-

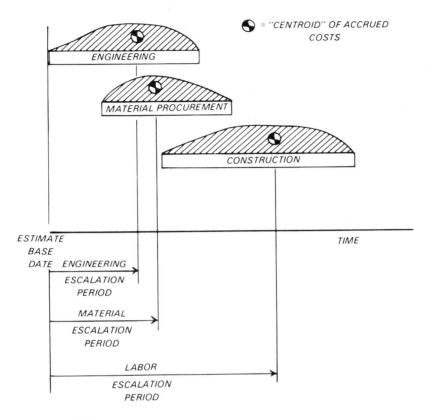

Figure 5.3 Calculating escalation for the typical small project.

eral, whereas "specific escalation" refers to that part of the market that affects the type of project under consideration. Frequently projects experience escalation trends that are markedly different from those of the economy as a whole, due to the effect of supply and demand on the part of the marketplace affecting projects.

Estimate accuracy: This is another frequently misused term. Estimate accuracy is the tolerance within which there is a specified probability that the actual cost will fall. To express accuracy we need two things: the tolerance band, i.e., the variation above and below the base (usually expressed as "plus or minus X%"), and we must also specify the probability that the final cost will be within the tolerance limits. When asked about the accuracy of an estimate, many estimators will say, "It's plus or minus 10%." This is, unfortunately, absolutely meaningless, as it defines only the tolerance limits. The only correct way to specify estimate accuracy is to say, for example, "The probability is 80% that the final cost will be within plus or minus 10% of the estimate." This is illustrated by Figure 5.4, which shows that, given a tolerance limit, two projects can have very different accuracies. Methods to determine estimate accuracy are discussed in "Estimating Contingency," later in this chapter and Chapter 10.

Estimating database: An estimating database is a collection of cost data reflecting a common time period and average past performance. The data is usually organized in tabular form, by cost code, and expressed relative to design variables (e.g., cost per hp) or quantities (e.g., cost per work-hour). Current cost data is used to keep the database up-to-date.

Estimating methodology: An estimating methodology is a set of guidelines, procedures and correlations used to prepare an estimate from the database. While the database provides specific data, the methodology provides generalized relationships between design and cost variables. The database is the basis of the methodology and the methodology is used to prepare estimates.

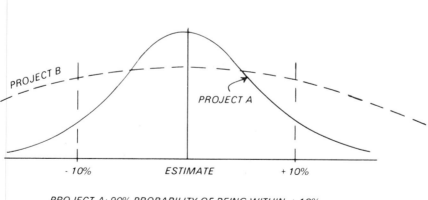

PROJECT A: 90% PROBABILITY OF BEING WITHIN ± 10%

PROJECT B: 60% PROBABILITY OF BEING WITHIN ± 10%

Figure 5.4   Defining estimate accuracy: The confidence limits within which there is a specified probability that the actual cost will fall.

ESTIMATING MATERIAL COSTS

Material costs are divided into two categories: equipment costs and bulk materials costs.

Equipment Costs

Vendor Quotes

Many companies estimate equipment costs by means of vendor quotes. There are two types of vendor quotes: a budget quotation, which is not binding and which is used solely for budget purposes, and a bid quotation that is binding and that appears on a purchase order.

Although budget quotes from vendors generally have a high degree of credibility, they are not necessarily a good source of estimating data. For instance, the item of equipment for which a budget quote has been given may be quite different from that which is eventually bought. The vendor will usually quote a standard, "off-the-shelf" item that may or may not meet all the specifications that will apply. By the time the design is completed and the specifications included, the actual item of equipment often turns out to be much more costly than that which was quoted originally. Also, the vendor has a lot of incentive to see the project go ahead and the order placed with his company. He can hardly be expected to be unbiased. In general, vendors will be inclined to give the lowest budget quote possible, although there may be cases, such as when they are to be the sole bidder for a special item, in which they may want the project leader to be conditioned to the idea of a high price. In order to compensate for the vendor's bias in budget quotes, a "vendor bias adjustment" is usually added. This adjustment is usually around 10%, but the actual amount will vary depending on conditions.

Even bid quotes need to be conditioned. Most purchase orders start with no revisions but end up with quite a few, most of which result in added cost. Therefore, even a bid quote needs to have a "developmental allowance" added to it, usually in the range of 3 to 8% depending on the complexity of the equipment involved and the likelihood of revisions.

Historical Data

Another way to estimate equipment costs is through the use of historical data. Such data may be collected by the company involved (see "Use of Historical Data," later in this chapter). Standard reference manuals are also available, as well as computerized estimating methods. Many equipment vendors will provide pricing books, formulae, and graphs so that a preferred client can do his own budget quotes. Historical data is often the preferred way to estimate as it reflects actual past experience and includes miscellaneous costs that might otherwise be overlooked.

Checklist

When estimating material costs it is important to recognize the various cost elements that make up the total cost to the project:

The *cost of the equipment* itself (usually the base quote): This must include all items necessary for complete installation.

Spare parts: these can be "warehouse spares" or "capital spares," which are high cost, special purpose spare parts (e.g., compressor rotors)

Special packing: some equipment requires special packing that may not be included in the price

Freight: These changes may vary as equipment may be quoted as "FOB" (freight-on-board), "FAS" (freight-along-side), or delivered to site.

Special shipping: This is for complex, fragile, or very large equipment

Vendor reps: These are technical representatives who assist in the installation and startup of a complex piece of equipment

Taxes: federal, state, and local

Duty: on imported items

Escalation: is the price fixed or is the price in effect at time of delivery?

Insurance

Exchange rate fluctuations (for imported equipment)

All the above items may contribute to equipment cost. The estimator should check that all have been accounted for in the equipment cost or elsewhere in the estimate.

Bulk Material Costs

Bulk materials, such as piping, electrical, structural steel, instrumentation, insulation, and paint, pose one of the most difficult problems in estimating, especially where small projects are concerned. In most cases, the actual quantities of each type of bulk material will not be known until long after the estimate is prepared, and, even then, the many different categories of materials, each of which has a different price, make preparation of an estimate a time-consuming task.

Small projects have the added problem of not having enough time for detailed estimating, as well as the difficulty of estimating modifications to existing facilities. This is due to the routing of new cables, pipes, etc. that, in an existing plant, may be determined by interference factors that are not evident in the design stage. So bulk materials are an area of likely overrun, and a realistic estimate is important to the success of the small project.

Data from Similar Projects

How then to estimate bulk materials for the small project? One way is to use data from similar past projects. When similar past projects are used, however, adjustments must be made to reflect the difference between the past project and the current one. For example, we might note that we used so many tons of piping on a similar project last year, and that this project should have about 10% more linear feet of piping, with an average line size 2″ greater. The calculation might look like this:

Past Project

Piping cost: $15,000
Piping quantity: 2000 linear ft., 8 in. avg. diam.
Piping complexity: average
Piping materials: carbon steel, schedule 40

Current Project

Piping quantity: 2200 linear ft., 10 in. avg. diam. (price is +21%)
Piping complexity: more flanges and fittings required
Piping materials: carbon steel, schedule 40

Piping cost calculation:

let B = base cost (from previous project)
  Q = ratio of quantities = 2200/2000
  S = ratio of avg. unit price for 10 in. piping/8 in. piping
  C = factor for complexity, more flanges and fittings: say 1.40
  E = escalation factor from then till now, say 1.12

then: piping cost = $B \times Q \times S \times C \times E$
      = (15000)(1.1)(1.21)(1.4)(1.12) = $31,305

Factors Applied to Equipment Cost

Another popular way to estimate bulk materials is with factors, or ratios that indicate the amount of bulk materials in each category that are typically associated with a certain type of equipment. For example, we might multiply the cost of a pump by various percentages to get the cost of the associated piping, electrical, and foundation work. These factors, which can be found in published literature on cost estimating, are often not appropriate for small projects. They reflect average experience, and the specific layout requirements for small, revamp-type projects are apt to be quite different. However, for those small projects that are repetitive, or those for which good data is available (e.g., architectural work), the factoring approach can work well.

Preliminary Takeoffs

A third approach is the use of preliminary "takeoffs," in which the quantities of bulk materials are derived from the measurement of typical distances, taken from drawings. For example, the plot plan might show the distance from some motor-driven equipment to the substation and to the control house. From these dimensions, an average cable-run length can be derived, and an estimate made of the total cable requirements. Even better, more detailed takeoffs may be available, such as those used for purchasing. An advantage of preliminary takeoffs is that they tend to produce estimates of the right magnitude most of the time. Although less exact than detailed takeoffs, they are less likely to miss major quantities than the factoring method or a detailed takeoff performed too early in the design process. Whenever takeoffs are used, a "takeoff allowance" must be included to cover those items that are likely to have been missed. A typical takeoff allowance can be 25% or more depending on the level of detail of the takeoff.

Once the takeoff is complete, the costs can be estimated by deriving an average unit cost based on past data or on information from vendors and reflecting the mix of sizes and materials that are expected.

ESTIMATING DIRECT LABOR COSTS

Use of the Network and Resource Plan

Like bulk materials, labor costs are apt to be particularly difficult to estimate for small projects. When the network has been completely resourced, we have an ideal basis for estimating the direct labor costs. The total work-hours (represented by the resources on the network multiplied by the activity durations) can then be multiplied by the appropriate labor rates to obtain the labor cost estimate. This process is straightforward and simple, and can be easily computerized as shown in Figure 5.5.

One advantage of this approach is that the estimator thinks in terms of the numbers of people required to do the work in an activity, rather than the number of work-hours. When estimating is being done by judgement, it is important to define a method that is based on a format for

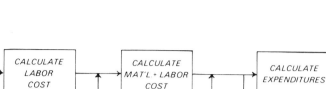

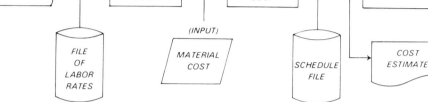

Figure 5.5 Computerized estimating of network resources.

judgement that matches the way people think, and many project people think more easily in terms of "how many people for how long" than in "how many work-hours."

Use of Quotes

For certain types of work, quotes can be obtained from the potential contractor. This practice can be effective in cases where the work involved is quite similar to previous work done by that contractor. However, use of such quotes should reflect consideration of bias, potential growth in the scope of work, and other aspects of the work that might not be reflected in the quote.

Use of Data from Similar Projects

Data from similar past projects can be analyzed and adjusted to derive a labor estimate for the current project. The reconciliation process should consider the relative complexity and difficulty of the two projects, the quantity of work involved, the type of labor, and the conditions under which it will be done. For example:

Past Project

Labor cost: $20,000
Direct work-hours: 960
Productivity: average

Current Project

Scope of work: greater than past project by 20% (using judgement)
Productivity: expected to be better than the past project due to:

Simpler work: 3% improvement
Summer weather: 2% improvement
Better access: 4% improvement

Wage rate: 15% higher due to escalation
Calculation of labor cost:

($20,000)(1.20)(0.97)(0.98)(0.96)(1.15) = $25,187

Use of Historical Data

Data on unit labor costs (e.g., work-hours per in. or ft. of pipe, work-hours per ton of structural steel, work-hours per cubic yd. of concrete) can be accumulated in a database and used for estimating. This is a preferred method in most cases, because actual data is generally the best predictor of what will happen in the future (if it happened that way on ten similar projects, chances are it will happen that way again). Such unit labor rates are also available in estimating manuals or computer databases. When using industry-standard data, however, it must be recognized that the data is representative of some selected locations and conditions, which may or may not reflect the performance of the project in question. Multipliers are generally developed by companies using these outside services to reflect local productivity. Once the work-hours are determined, the hourly rates prevailing at the project location can be applied to get total labor cost.

Use of Factors Applied to Equipment Costs

Labor costs can be estimated in the same fashion as bulk materials, by using factors or percentages of the equipment cost. As in bulk-material estimating, this can be a dangerous practice for small projects as the relationship between equipment costs and labor for equipment and bulk material installation may be quite different for the particular small project than for the average project on which the factors are based.

Components of the Hourly Labor Rate

When estimating man-hours and costs, an hourly labor rate is used, and it is important for the estimator to know what is included and excluded. Hourly labor costs are made up of some or all of the following components:

Direct wages
Taxes (e.g., employer's contribution to F.I.C.A., unemployment tax, etc.)
Holidays
Sick days
Vacation
Overtime premiums
Insurance premiums (for life and health insurance)
Training
Overheads (consisting of supervision, transportation, supplies, support
   staff, administration, support facilities, and light contruction equip-
   ment)
Profit (if labor supplied by a contractor)

The estimator should know whether or not these items are included in
the hourly rate and, if not, where they are covered in the estimate.

Productivity

Estimates of labor work-hours must, either explicitly or implicity, contain an assessment of labor productivity, i.e., the number of work-hours
required to accomplish a fixed scope of work. Productivity is generally
defined as the *planned work-hours expressed as a percentage of actual
work-hours required to complete a given scope of work*:

$$\text{Productivity} = \frac{\text{planned work-hours}}{\text{actual work-hours}}$$

Using this definition, productivity over 100% is "good."
   Productivity is apt to be different from job to job, even at the same
location, so it is important to make any adjustments in the labor work-
hours that might be necessary to account for productivity variations for
every estimate.
   Some of the factors that affect direct labor productivity are:

Job size: Small projects can often achieve levels of productivity that are
   orders of magnitude greater than large projects in the same location.
   This is due to their ability to utilize small, well-organized subcontractors, to supervise closely, and to obtain a small, high-caliber
   workforce.
Job complexity: Small projects involving complex equipment, sophisticated electronics, or unusually difficult construction practices (e.g.,
   excessive work above- and below-grade, extreme congestion, etc.)
   are likely to experience lower productivity.

Market activity: Periods in which contractors are looking for work are often characterized by high productivity, since the best workers can be made available and contractual incentives for good performance can be negotiated.

Type of workforce; whether union or nonunion: If the data used to estimate work-hours is based on a union (or a nonunion) workforce, and the job at hand uses nonunion (or union) labor, the productivity may vary.

Type of workforce; whether in-plant or contractor: The plant forces are likely to have different productivities than the contractor forces, and the type of workforce to be used should match the data used in the estimate.

Schedule duration relative to total manhours; for a given project, there is an optimum schedule duration. If we have a tight schedule, increased manpower concentrations will be necessary, with the attendant loss of supervision, increased interference, and reduced productivity. Similarly, a longer-than-optimum schedule can reduce productivity by diluting supervision, as well as by loss of momentum and continuity.

Shift work: Productivity is generally lower on the evening shift than on the day shift, and lower still at night. These productivity reductions are due to the need for coordination of work between shifts, darkness, and reduced supervision. The evening-shift productivity is perhaps 80% of day-shift productivity, and the night shift is around 60% of day-shift productivity. These figures will, of course, vary greatly between projects depending on working conditions and the effectiveness of the planning and supervision.

Overtime work: Use of overtime generally results in lower productivity due to the problems of maintaining work efficiency over longer periods. There are some cases in which overtime is considered a financial bonus and therefore little extra work is accomplished.

Changes: Field changes can be due to design changes, rework of fabricated equipment or bulk materials, rework of construction work, or a change of the work scope as the project progresses. These changes tend to interrupt the flow of work, and often require work that is inherently more difficult than the original scope. They therefore tend to reduce productivity.

Based on consideration of all of the above, the estimator should make whatever adjustments are necessary to reflect the productivity expected for the project. These adjustments can be judgements, but they should be clearly documented. For example:

Base productivity from reference data:		90%
Adjust for: project size (smaller than normal)	+5%	
work complexity (higher than normal)	(10%)	
market conditions (normal)	n/a	
type of workforce (plant forces)	(10%)	
tight schedule	(8%)	
shift work (20% of work at 15% less productivity)	(3%)	
changes (normal)	n/a	
Productivity expected for current project:		
(90%)(1.05)(0.9)(0.9)(0.92)(0.97) =		68%

ESTIMATING FIELD OVERHEADS

Field overheads, which include supervision, indirect labor, temporary facilities, and construction equipment, can be an important part of the small project. Where the work involved is complex, the field overheads can be a major part of the cost. There basically are two ways to estimate field overheads: using a detailed method, or as a percentage of direct labor.

Using a Detailed Method

The detailed method is preferred for many small projects because it enables the estimator to use the particular requirements for the project at hand, which may vary substantially from previous projects. For example, a small project may require a large mobile crane to install a certain item of equipment. The cost of this crane might represent a significant percentage of the project cost, and an abnormally high percentage of direct labor. Small, complex projects might also require an abnormal amount of supervision. The detailed method, then, requires the estimator to specify and price each category of field overheads. One of the best ways to do this is by using the planning network and loading the resources for field indirects (i.e., indirect field costs) onto each activity.

The principle of resource planning is that *the resources required to accomplish the given scope of work in the given duration must be provided if the work is to be accomplished.* These resources can include major items of field overheads such as construction equipment. Other field overheads, which are not assignable to specific activities, can be estimated in detail or as a percentage of direct labor. The field overheads on a small project may include:

Indirect labor: This includes the timekeeper, accountant, payroll clerk, guard, purchasing agent, warehouseman, equipment operator, driver, x-ray technician, surveyor, mechanic, clerical and office staff, site cleanup crew, and janitor.

Supervision: Including the foreman, planner, inspector, site manager, construction superintendent, safety supervisor, craft supervisor, field engineer, and subcontract administrator.

Temporary construction: This may include the field office, warehouse for construction material, equipment shed, pipe fabrication shop, other fabrication facilities, change houses, field toilets, guard house, cafeteria, first-aid facility, and laboratory. The furnishings and equipment for these temporary buildings must also be considered as well as the temporary power, water and sewer lines, and phone connections. Roads, ditches, parking areas, staging areas, fences, and signs are also temporary facilities that may be required. Note that all the above refers only to temporary construction, i.e., that which will be removed at the end of the job. Permanent facilities are part of the workscope and are not included in this account.

Consumables: Including welding rod, gloves, protective clothing, small supplies (e.g., nuts, bolts, gaskets, nails, and clamps), lubricants, fuels for equipment, hardhats, ropes, rags, etc.

Construction equipment: Including lifting equipment (e.g., cranes, gin poles, derricks, and hoists) automative (e.g., cars, and trucks of various types), earthmoving (e.g., backhoe, bulldozer, front-end loader, roller, and excavator), cranes (e.g., "cherry-picker" mobile crane, crawler cranes, and sideboom tractor), welding equipment (e.g., welding sets and stress-relieving kits), shop equipment, testing equipment, communications (e.g., radio and telemetry), special-purpose equipment, and miscellaneous small equipment (e.g., pumps, stirrers, hammers, compressors, blowers, ladders, scaffolding, etc.).

Calculating Field Overheads as a Percentage of Direct Labor

This approach is often used for large projects, in which the field indirects are fairly consistent. It is appropriate for small projects that are reasonably consistent in terms of size and complexity. Since certain field overheads are required regardless of job size, the overhead percentage can be highly variable, and the estimator is cautioned to be sure that the data on which the percentage is based is consistent with the project at hand. If not, an adjustment can be derived from past data by developing a curve of field overhead percentage as a function of total man-hours.

INTEGRATED COST ESTIMATING

Many companies have, in recent years, formalized the process of estimating in a way that integrates cost, time, and resources. In fact, this practice has become sufficiently accepted that most project-management software uses this approach.

The integrated approach to estimating begins with the project plan as shown on the management network. For each activity, the resources required and the total cost are estimated (see Figure 2.6).

The integrated method has found rapid acceptance for small projects because it clearly shows the relationship between cost, time, and resources, and is therefore more flexible than traditional methods and better able to show how costs are affected by variations in the project execution. This format is particularly attractive for many small-project applications because of its simplicity, logic, and flexibility, and it provides an excellent basis for project control (see Part III).

ESTIMATING ENGINEERING COSTS

Engineering costs are estimated in a manner similar to that used to estimate direct labor costs. Acceptable methods use the resourced network plan, quotes from engineering contractors, data from similar projects, and historical data.

Use of the Network and Resource Plan

If the network has been resourced for engineering work, we have an excellent basis for estimating engineering costs. Using the same procedure as for labor costs, the number of people in each category × the duration of the activity × the appropriate man-hour rate will give the cost estimate, as illustrated in Figure 5.5.

Use of Quotes

Budget quotes for engineering work are difficult to obtain, since engineering work is often done on a reimbursable cost basis. However, in small projects where the amount of engineering work is readily defined, it may be possible to get a lump-sum budget quote.

Use of Data from Similar Projects

Data from past projects is often a good indicator of the engineering content of a project. The data should, of course, be adjusted to reflect the scope and complexity of the project at hand. This adjustment is often made by calculating the engineering costs as a percentage of direct material and labor cost.

Detailed Estimate Based on Scope of Work

Engineering work can be estimated in detail by estimating the number of drawings and specifications required based on the number of equipment items and piping lines contained in the basic design. This approach is not generally suited to the small project, as the necessary data and methods are usually not available. However, in some cases it is possible to calculate the number of drawings and specifications and multiply by the estimated hours per drawing to yield the total design manhours. Other costs can then be estimated in detail or calculated as a percentage of the design work.

Components of the Hourly Engineering Rate

As engineering costs are usually handled in terms of work-hours and cost per work-hour, it is necessary to know what is included and what might be excluded in the rate being used. Those necessary items that have been excluded should be covered elsewhere in the estimate. Hourly rates for engineering are made up of some or all of the following components:

Salary
Taxes (e.g., employer's contribution to F.I.C.A., unemployment tax, etc.)
Holidays
Sick days
Vacation
Overtime premiums
Insurance premiums (e.g., life and health insurance)
Recruiting
Training
Overheads (e.g., supervision, office facilities (floor space, utilities, furnishings, etc.), computer facilities, office staff and services (secretarial, mail, telex, telephone, janitorial, and cafeteria), marketing expenses, cost of financing, company liability insurance, recruiting,

internal R & D, library and databank, specifications, procedures, etc.)
Profit (if engineering work provided by a contractor)

Productivity

Engineering, being a form of labor, will experience productivity variations under differing conditions. As in labor productivity, engineering work is done far more efficiently during periods of low contractor activity. During these periods, contractors are likely to cut back on staff—so that the best people are available to work on the project—and more favorable contract conditions can be negotiated to help assure good performance.

Job size and complexity will also affect engineering productivity, with smaller and simpler projects capable of higher productivity. However, in an engineering operation, the small project is apt to get far less attention, perhaps even lower priority than a large one, and therefore the productivity may well be less.

Schedule and job size affect engineering productivity in the same way as they do field labor. Tight schedules foster low productivity as progress becomes more important than efficiency. In addition, the use of overtime often leads to significantly higher costs and lower productivity. Engineering work, particularly on a small project, is apt to be affected strongly by the quality and characteristics of the people involved. One should make an extra effort to identify those individuals who are most likely to do a good job, and have them assigned to the project. (This is also true for key personnel on the field supervision staff.)

ESTIMATING MISCELLANEOUS ITEMS

No estimate is complete without consideration of the various miscellaneous items that affect all projects. Typical miscellaneous items include:

Taxes (federal, state, and local)
Fees (for special operations such as large transport over city streets)
Insurance (for materials, personnel, liability, etc.)
Permits
Interest charges
Exchange rate fluctuations
Legal assistance

These miscellaneous costs can be estimated in whatever way is most suitable. The important thing is to be sure all miscellaneous costs are covered in the estimate.

ESTIMATING ESCALATION

Most small projects involve a relatively short time period, so escalation is not as big a problem as it is for larger projects. However, the elapsed time from the base date of the estimate (i.e., the time period represented by the data used for estimating) to the time at which the work will actually be done can be significant. Most small project cost estimates must, therefore, make some provision for escalation.

Estimate Components

The calculation of escalation costs usually begins by breaking the estimate into components that are characterized by distinct escalation rates. The estimate components that tend to have distinct escalation trends are:

Direct materials (equipment and bulk materials)
Direct labor
Field overheads
Engineering services
All other costs

Escalation Indices

An escalation rate is then established for each cost category. This rate may be calculated a number of ways but it must be remembered that we are looking for the specific escalation that relates to the project, not escalation rates for consumer or other unrelated goods. The sources of data and indices for tracking and forecasting escalation include:

Federal *government indices* (e.g., Bureau of Labor Statistics)
Escalation indices developed by *industry groups* (e.g., labor unions)
Escalation indices published in *professional journals* (e.g., *Oil and Gas Journal*'s "Nelson Cost Index")
Historical data from company files or database
Economic reports by leading banks, universities, and industry groups
Guidelines and forecasts published by the *financial department* of the company (used for forecasting, budgeting, financial planning, etc.)
Services provided by *consulting firms* to industry

For small projects, it is best to develop and agree which indices are to be used, update the indices periodically, and use them on all projects.

Calculating Escalation

To calculate the escalation costs, the project schedule is used to establish the centroid of the expenditure of each category of project cost, i.e., the point in time at which, if we purchased everything in that cost category at once, our cost would be the same as the actual condition in which we make purchases over time (see Figure 5.3). The length of time from the base estimate date to the centroid is the time period for which escalation is to be calculated. This amount of time, multiplied by the escalation rate per unit of time, gives the cost of escalation for that category. If the estimate is to be used for cost control, the escalation must be factored into each item in the estimate, so that the escalation account disappears and each cost item is shown as "money of the day," that is, the expected cost at the time the money is spent.

ESTIMATING CONTINGENCY

Estimates Must Include Contingency

One item that is included in all estimates is uncertainty. Since an estimate is, as we have said, a prediction, it is inevitable that the project actually develops in many different ways from what we expected. In most cases, it will turn out to be different in a way that results in increased cost. To bring the cost estimate up to a level that has a reasonable chance of not overrunning, we add contingency.

Contingency is a much misunderstood cost engineering concept. It is also the part of cost estimating for which many unsatisfactory estimating methods exist. Small projects have a risk of overrunning—in many cases the risk is greater than for large projects—so contingency must be included in the estimate to cover design changes, variations to the planning basis, and estimating differences with actual costs. Chapter 6 is devoted to this subject and describes a number of methods for estimating contingency. For convenience, some of the methods are summarized below.

Contingency calculations are particularly important for cost estimates that are to be used for bidding purposes, in which we wish to study the relationships between the bid price and the probability of winning the job.

Summary of Methods for
Estimating Contingency

Judgement, Based on the Risks Foreseen for the Project at Hand

This commonly used method can actually provide reasonable results in the small project environment, but it suffers from being purely qualitative. Studies have shown that project leaders and managers tend to be unduly optimistic about the chances for success. And, there may be risks that are overlooked in such an analysis. Sometimes, contingency assessments are made by committee which, depending on the group dynamics involved, may give a better (or at least a more acceptable) result.

Guidelines or Procedures Derived from *Historical Data*

Many companies set contingency by policy, such that all estimates submitted for approval are given the same contingency. This method, probably the most commonly used, ignores the fact that some projects have more uncertainty or risk than others and therefore require more contingency. Those high-risk projects are likely to have contingency "buried" in the estimate, thereby fostering a sense of confusion and mistrust, as well as diminishing the estimate's usefulness as a control tool. Another problem with this approach is that historical data, i.e., what *has* happened, is not always the best predictor of what *will* happen.

Comparison with Past Projects Having Similar Risks

This method can give reasonable results for the small project, if the selection of projects for analysis and comparison is made carefully, and the analysis done objectively. For example, the project leader in this case might say "We had 20% contingency on a similar project last year, and needed every bit of it. However, the current design reflects certain lessons learned from that project, and so the portion of contingency that covered those unforeseen changes will not be needed on this project and we can use a contingency of 17%. This project also is in a more difficult area of the plant to work in, but we had better data to use in our estimate." This thought process can be documented as follows:

Contingency provided on similar past project:	20%
Final cost as % original estimate:	123%
Portion of overrun due to contingency variations:	1%
Thus, contingency used on past project:	21%

Differences in estimating bases between past and present projects:

Design: Current project has less potential for changes
Planning: Current project is in a more congested work area
Cost: Current estimate used better data

Contingency required for current project:

Design variations: say 3% less than past project
Planning variations: say 5% more than past project
Estimating variations: say 6% less than past project

Net contingency requirement = 21% − 3% + 5% − 6% = 17%

Use of a *"50/50 Project Model"*

In this and the preceding chapters, we have seen how a small project is planned, scheduled, resourced and estimated using the integrated technique that results in the definition of the "project model." The original project model is comparable to the base estimate in that it contains no contingency for cost or schedule variations. Therefore, due to the inherent skewness in project data described previously, the base model is more like a "target," that is, it represents a cost and schedule performance that is worth trying for but which we do not expect to achieve. So, the probability of the "target model" being achieved or bettered is much less than 50%.

It is possible, then, to create a "50/50 model" in which the durations and costs reflect the variations covered by contingency, and the probability that the cost and schedule will be met or bettered is 50%. This model can be compared with the target model and, the differences considered to be contingency.

This approach is well-suited to small projects, as it shows clearly the cost and schedule effects of the expected variations, it is easy to do, it avoids use of statistical concepts and terminology with which (unfortunately) few people are comfortable, and it can be easily related to past data.

Manual Calculations Using Probabilities

A number of methods are available for calculating contingency by identifying specific risks, assessing their probability and cost impact, and performing probabilistic calculations. The best known methods in this category are "decision tree" analysis, which uses a logic diagram, and "expected value" calculations which use a tabular presentation. Although one or both of these methods can be useful in certain situations, because they deal with discrete risks they are better suited for risk analysis and decision-making than for contingency calculations. There are a number of excellent texts on this subject for the reader who wishes to try these methods.

Probabilistic Simulation

There are a number of computer programs available that do a credible job of calculating contingency using a technique known as "Monte Carlo Simulation." These programs may be suitable for some small projects in which a detailed and rigorous approach to risk analysis and contingency calculation is required. The simulation technique attempts to define the probability function for the project cost by simulating a large number of possible cost outcomes, using, as input, probability functions for each of the cost variables. This technique is widely used in many other applications, and is well-accepted. Its disadvantage in the project environment is that the cost variables must be independent for the method to work properly, and, in most cases, this is not the case. The reader desiring more information on this technique is again referred to the many excellent books on risk analysis, operations research and decision theory that describe it thoroughly.

ESTIMATE PRESENTATION

Once the estimate has been completed we are ready to summarize and prepare it for presentation and review. This part of the estimating process often proves to be just as important as the estimate itself, for, if the estimate lacks credibility, it will not be accepted by management, and the project may not be approved. If this happens, the estimate will be useless as a control tool. How then can an estimate be documented and presented so as to assure it's credibility and usefulness?

Defining the Estimate Basis

The most important aspect of estimate documentation is not the estimate itself, but its basis. The "tripod" of the design, planning, and cost bases is what supports the estimate, and it can be no better than that base. Thus, the estimate basis must be clearly documented.

The design basis: This includes specific drawings, sketches, and specifications; date of release, revision number, and person(s) responsible; important design assumptions; overall scope of work understood to be involved; specific inclusions and exclusions of facilities to be installed; and work to be done.

The planning basis: This involves the network plan and schedule, resource plan, and contracting plan; date of release, revision number, and person(s) responsible; and important assumptions as to how the work will be executed (e.g., hours worked per week, use of overtime, contractor performance expected, etc.).

The cost basis: This involves the estimating method(s) used (and why they are appropriate for the project); sources of estimating data; cost level of the base estimate (i.e., what point in time is reflected); productivity and hourly cost rates used for engineering and labor; and escalation rates used.

The Estimate Summary

The estimate summary should provide a one-page summary of the costs, generally organized as follows:

Direct material
  equipment
  bulk materials
Direct labor
Total direct costs
Field overheads
  supervision
  indirect labor
  construction equipment
  temporary facilities
Engineering
  in-house
  contracted
Miscellaneous
Escalation (if not included above)
Total base estimate cost
Contingency
Total cost estimate

The estimate summary can then be further broken down by areas, cost centers, or according to any other practice followed by the company.

A final small but important point: cost figures should be rounded; nothing makes a worse impression than precise estimate figures (i.e., to the nearest dollar or penny) that imply an estimating accuracy that clearly is not possible.

Reconciliation to Past Estimates

The reconciliation is a very important aspect of estimate presentation that is often overlooked. Just about every estimate has some sort of estimate that preceeded it, and every manager, looking at the new estimate, will mentally compare it with the old. Credibility is characterized by clarity and thoroughness (but without excessive detail), as well as by having the answers to unasked questions. The reconciliation provides all these characteristics. The reconciliation should be rigorous but easy to understand. Large reconciliation items can be explained with footnotes as needed. The method for preparing reconciliation is described in "Reconciling Forecasts and Estimates," earlier in this chapter.

Estimate Sensitivities

In some companies, it is desirable to present to management the specific areas that the estimator feels are particularly likely to cause an overrun. In many cases, the credibility of the estimate is enhanced by an open, frank, and well-thought-out presentation of the specific risks and sensitivities involved (note that these are specific risks that are not included in the contingency analysis). For this type of manager, a page showing the sensitivity of the estimate to various risks is often appreciated. For example:

Total cost estimate	$350,000
Possible scope change (new road)	+ 55,000
Possible scope change (rework stack)	+ 15,000
Possible strike (assume three weeks)	+ 85,000
Overseas purchasing (compressor)	− 30,000

It is tempting to add up to the plusses and minuses to get an idea of the total possible cost. This should never be done, as the laws of probability indicate that the probability of all these things happening together is the product of each individual probability and is therefore almost always infinitesimally small (see Chapter 3).

Estimate Details

The estimate details are unlikely to be reviewed by a manager, but they are important for building credibility with those who are doing the work.

The format for the estimate details varies widely from company to company. What is important is that the details be clear, explicit, and appropriate for their intended use.

The intended use of the estimate details is cost control. Therefore, the level of detail in the estimate summary should be comparable to that of those who will be using the estimate. Some estimates that have used adjustments from similar projects or other gross estimating techniques contain too little detail for control and this should be recognized before attempting to use them for that purpose. In those cases, detailed data from past projects can be used to establish a breakdown of costs for the control estimate.

REVIEWING AN ESTIMATE PREPARED BY OTHERS

Project leaders often find themselves in the position of reviewing and either accepting or rejecting an estimate prepared by someone else. The estimate may have been prepared by a subordinate, by a contractor, by the estimating department, or by another project leader. Those who have had to review estimates have probably found that it is not very easy. What questions should I ask? What should I look for? How can I be sure that the estimate is as good as one I did myself? These questions are important because the estimate reviewer will probably have to approve it, and therefore share in the responsibility for its being correct. This problem of estimate reviews is often made much worse by the fact that the time left for reviewing is usually very short by the time the estimate is finished, and no one wants to be accused of holding up the project.

Objectives of the Estimate Review

An estimate review can be thought of as a series of questions and checks, designed to accomplish the following reviews and answers the following questions: Fortunately, there are some guidelines that can be followed.

1. *Evaluating the basis of the estimate*: It is complete and is it likely to change? Is there any new technology involved? If so, how was that reflected in the estimate? Are the plan and schedule well-thought-out? Is the schedule a 50/50 case? If not, have schedule premiums been included? Are there are unusual aspects to the work? If so, how have they been reflected? How is this project similar to previous projects? How is it different? Have the differences been accounted for? Are the assumptions on which the estimate is based clearly stated and reasonable?

2. *Reviewing the methods and data used*: Are the estimating methods appropriate to a project at this stage? Have these methods given good results on previous projects? What parts of the project were difficult to estimate? How were they handled? What was the source of estimating data? Is it representative of conditions anticipated for this project? Does it have the correct amount of detail? Has it been adjusted where necessary to reflect this project? Is it current? Does it exclude any elements that should be in contingency? Was a computer used for estimating? If so, how was it used? Was the output verified?

3. *Assess the caliber of the personnel involved*: Are they experienced in estimating? Are they aware of the parts of the estimate most subject to variation? Have they spent enough time and effort on the critical estimating variables? Did they check their work? Was it properly reviewed?

4. *Evaluate the quality of documentation*: Can every number on the summary pages be traced back to the backup? Is it clear how each number was calculated? Is the estimate in a format suitable for cost control? Is it clear that the estimate has not been "padded" in any way? Have all assumptions been clearly documented?

5. *Check the absolute value of the estimate*: Is the estimated cost consistent with other similar projects? Are the relative values of the major cost components about what one would expect? What special features does this project have which might make it difficult to base the estimate on past experience? Is there any reason to expect that the estimate is biased high or low?

Techniques for Estimate Review

The techniques for estimate review are based on these principles:

Avoid Getting Involved in the Details Unless
Absolutely Necessary

Many of the mistakes made by estimators and project leaders happen because they are so swamped by details that it becomes impossible to draw any conclusions. Many find themselves so busy examining the "bark on the trees," that they never get to realize that the "forest" is blazing away behind them. Those who review an estimate by checking the arithmetic (not an uncommon practice) are good examples of this practice that can be improved by refusing to get involved in details unless absolutely necessary. Be aware that this may be difficult, as the "reviewee" may feel safe by keeping the reviewer bogged down in detail.

Remember the "80/20 Rule"

This is also a favorite of auditors and consultants, also called "Pareto's Law," the "ABC Rule," and the "Law of the Significant Few and Trivial Many." It simply states that there are, in any situation, a few items or aspects (say 20%) that make most (say 80%) of the difference. This is definitely true for cost estimates. Almost inevitably, the cost variables that have the greatest impact on project cost, and that vary the most are:

Labor productivity
Bulk material quantities
Unit work-hours for labor
Engineering workhours
Hourly rates for labor and engineering
The scope of work
Escalation (if applicable)
Design changes
Market conditions
Inclusions and exclusions
Contingency

In general, if these aspects of the estimate have been handled correctly, there is not a lot that can be wrong with it. So, it is wise to spend most of the time for estimate review on these items which, if wrong, have the greatest impact.

Using the Auditor's Questioning Method

Auditors are skilled at asking questions that will result in an answer that provides a good indication of whether more questions need to be asked. For example, a satisfactory answer to question 1 means that we can proceed to question 2 whereas an unsatisfactory answer means that there may be some weakness in that area and we better explore it more deeply (with question 1a). Skilled managers who seem to have the uncanny ability to, in five minutes, pinpoint the one mistake in a month's work, are probably using this technique. This approach works very well in estimate reviews, since there clearly isn't time to go through the entire estimate. It depends, of course, on the questioner's ability to be unimpressed by charm and by being told what one supposedly wants to hear.

Use the Auditor's Sampling Method

Auditors also like to sample; to take an item at random and thoroughly check everything there is to check about it. If everything holds up well, chances are the other parts of the estimate are reliable also.

Use the "Vertical Audit" Technique

This technique is used most often in conjunction with sampling and questioning. It involves following, in detail, a procedure that is typical of the work done. For an estimate review, it means examining every step of the estimating process. The following conversations illustrate the concepts discussed so far. They might take place between the estimate reviewer and the person responsible for the estimate:

Q: "Where did these electrical costs come from?"

A. "The electrical department. Those guys really know their stuff. In 30 years, they've never been wrong on an estimate. You can be sure those electrical costs will be right on the money!"

Q: "How was the estimate done?"

A: "The same way they do all their estimates. Why should they change when they've been so successful?"

Q: "How was it documented? Where is the backup to the estimate? Where are the quantities? How were they derived? On what drawings were they based? Were those drawings up-to-date with the latest changes? What is the basis of the cost data? Is it relevant to this project? Is it consistent with the rest of the estimate? What have you done to make sure that it includes everything that should be included? Who did the electrical estimate? Who checked it? When was it done?"

A: "Uh, we better go talk to the electrical department."

In the electrical department, of course, a similar conversation takes place, going, if necessary, into progressively greater levels of detail until the reviewer is satisfied that the estimate is sound or has concluded that it is unacceptable. Compare that conversation with this one:

Q: "Where did these electrical costs come from?"

A: "The electrical costs were prepared by the electrical department as they have the special knowledge and up-to-date data which is needed."

Q: "How was the estimate done?"

A: "The scope of work to be estimated by the electrical department was defined by the estimating department, along with the method to be used. The design quantities were derived from the current single-line drawings, equipment specifications, and plot plans. The cost data was taken from our current purchasing agreements with suppliers of cable, conduit, and fittings, as well as from budget quotes from two potential suppliers of the transformers and switchgear."

Q: "How was it documented?"

A: "The electrical estimate was documented in this package which was transmitted to estimating, reviewed for consistency, and then incorporated in the project estimate."

Q: "Who is responsible for the Electrical Department's work?"

A: "Bob Kable."

Q: "OK, let's call in Bob Kable and we'll all have a look at this package."

Note that the confident reviewee in the second conversation may only be telling us what we want to hear (though it sure sounds good), so we must make absolutely sure by insisting on going through the written details anyway. By involving Bob Kable, we make sure that we get an assessment of the people who actually did that part of the work. In any case, if the electrical estimate turns out to be sound, we can go through the rest of the project estimate with a more cursory review.

Use a Quick, Independent Check Estimate

There is no better way to build confidence in an estimate than to be able to arrive at the same answer in a different way. Even if one is reviewing a detailed estimate, it is almost always possible to check it by coming up with a "quick and dirty" check estimate. Techniques for doing the check estimate include adjustments to similar past projects, rule-of-thumb estimates, and curve-type estimates. If one can show that the estimate being reviewed is "in the same ballpark" as a rough, independent estimate, chances are it is reasonable. If not, of course the check estimate may also be wrong, but it indicates that some time should be taken for the estimate review.

EXPENDITURE FORECASTING

As part of the budget package the project leader generally must prepare an expenditure forecast to enable the financial department to assure that the necessary funds will be available when needed. This part of the estimating process is notorious for being difficult and unsatisfying as it tends to be a time-consuming task that generally produces erroneous results. The project model, containing both the cost and schedule of an activity, provides us with some better ways to forecast expenditures. In this context, we use the term "expenditure" to mean the actual passing of funds from one company to another. The "expenditure forecast" shows how much will be expended each week or month.

It should be noted that many companies forecast expenditures by using standard curves similar to S curves. However, just as the traditional

S curve is a poor predictor of small-project man-hour and progress trends (see "Details of Resource Planning," in Chapter 4), so too it is apt to falsely predict small-project expenditures. The techniques described below are easy to implement and will generally produce better results.

General Principles of Expenditure Forecasting

Step 1: Calculate the Expenditure Timing

Expenditure calculations must be done recognizing that the expenditure pattern will be different for different types of costs:

Labor costs (engineering, direct and indirect labor): If the work is done on a reimbursable basis, an invoice is usually submitted monthly for work during the previous month so labor expenditures tend to be linear over time.

Overhead costs (field and office overheads): These costs will also generally follow a linear expenditure pattern over time. They may be directly reimbursed, or included in the hourly rate.

Material costs: Materials are usually invoiced when delivered, fabricated materials may require progress payments. The expenditure pattern for materials tends to be more like a point or series of points.

In all cases, it is reasonable to assume that there is an elapsed time between the receipt of an invoice and the payment. This is due to the normal corporate practice of checking and approving the invoices, as well as holding on to company funds as long as possible. In most cases, a 30-day period is to be expected, but the accounting department is the best source for information on each company's practices. This lag in payment should be allowed for in the forecast. A general formula for calculating the timing of an expenditure is:

Date work is completed + time to prepare and submit invoice + time to pay

The project's schedule is, of course, the primary tool for determining the timing of the major payments.

Step 2: Calculate the Expenditure Amounts

The estimate and the contract plan are the basis for calculating expenditure amounts. The estimate allows us to group costs into major contracts (if this has not already been done). The type of contract will then determine the specific timing:

Lump sum: This kind of contract may give progress payments, or may be paid in full at the end of the work

Reimbursable: These may be owner-financed (in which case payments are made in advance) or contractor-financed (payments made monthly for work performed). May have a retention (usually about 10%)
Bonus: This may be included and is usually paid at the end of work

Using the Project Model for Expenditure Forecasting

Having prepared a resource- and cost-loaded network we know, from the schedule, when each activity will take place and how much each will cost. It is, therefore, straightforward to calculate how large the expenditures will be and when they will be required.

Resource aggregation provides an indication of the man-hours spent in each time period. If the estimate has been done on this basis we also know the cost of the invoice that will be submitted. By inserting the timing factor, we can easily generate the expenditure forecast. If the basic plan and cost estimate have been done with a computer-assisted method, the expenditure forecast can also be easily automated.

Using the Escalation Calculation to Forecast Expenditures

Since the escalation calculation requires us to segregate the estimate into categories for escalation that are similar to those for expenditure forecasting, and to locate those cost centroids in time, we can also use that calculation for expenditure forecasting. This is the process of "spreading" the escalated costs about the centroids used for escalation.

APPROXIMATE ESTIMATING

Estimators frequently encounter the situation in which a cost estimate is required immediately, must be prepared for a project about which little is known, and must be prepared in the absence of estimating data and methods. In other words, a situation requiring an approximate estimate. Approximate estimates, though not very accurate, have a number of important uses:

As a basis for decision making very early in a project's life
As a check on an estimate or bid prepared by someone else
As an interim estimate to justify funding and for further studies

There are two basic techniques for approximate estimating:

Adjustments to past projects
The "sixth-tenths rule"

Each of these is discussed below.

Estimating by Adjustments to Past Projects

This method uses the same technique described for reconciliations and in the estimating of bulk materials; that is, one should start with a known cost data-point, and make adjustments to reflect differences between the past project and the current project in terms of the design, planning, and cost basis. In some cases, it may be appropriate to make these adjustments using the six-tenths rule that is discussed in the next section. The technique is best illustrated by the example below.

Known cost of past project:		$100,000
Adjust for:		
Design variations		+ $2,000
equipment sizing	+ $3,000	
bulk materials, size and quantity	+ $2,000	
type of materials	($3,000)	
Planning variations		($1,000)
schedule duration	+ $4,000	
type of contract	($2,000)	
shift work	($3,000)	
Cost variations		+ $12,000
escalation	+ $12,000	
Cost of current project:		$113,000

In preparing approximate estimates it is important to clearly document the judgements, assumptions, and calculations made, no matter how approximate they are. In this way, one can always explain how the numbers were derived and, if the guesswork is criticized, the critic can always be invited to provide a better guess!

The Six-Tenths Rule

Cost engineers have observed over the years that, in many cases, the cost of something does not increase proportionately with its capacity. For example, if we wish to build a storage tank with twice the capacity of an existing tank, we will probably find that the bigger tank will not cost twice as much. Many people refer to this phenomenon as *"economy of scale."* In this case, we can easily deduce why economies of scale are in evidence. If the volume is doubled, the surface area of the tank will only increase 40%, and, since the surface area determines the weight that in turn determines the cost, the cost is likely to increase by only about 40% as well. For more complex facilities, there are many cost components that do not increase in a one-for-one manner with capacity. For example, instrumentation costs do not, in general, double when the size of the equipment is doubled.

When cost vs. capacity data is plotted, it often tends to take the form:

$$\left(\frac{\text{cost of A}}{\text{cost of B}}\right) = \left(\frac{\text{capacity of A}}{\text{capacity of B}}\right)^n$$

where n is typically about 0.6 (although it will vary).

This relationship appears, handily, as a straight line on logarithmic paper.

Different cost categories will have different values of n (found in various references), and the reader is cautioned not to extrapolate too far using the above relationship. However, the six-tenths rule is a handy way to make adjustments for approximate estimating.

CHAPTER SUMMARY

In this chapter we discussed the basic concepts and techniques of cost estimating, and demonstrated various ways to apply them to small projects. Having taken our hypothetical small project through the steps of network planning, scheduling, resource planning, and cost estimating we now have a complete project model. As we seek project approval, we can advise management about exactly what we intend to do (the design basis), how we intend to do it (the resource and contracting plans), how much it will cost (the estimate), and when we'll need the money (expenditure forecast). That is certainly an acceptable basis for the important process of project approval.

But the project model is a lot more important than just a means of getting the project approved. It serves as a basis for communication, commitment, and control—clear communication on our part as to what we understand our job to be and how we intend to do it, firm commitments on the part of those who must supply the resources we need, and effective control of our own work and that of others as the project progresses.

The project model will be especially useful for combining this project's resource requirements with those of other projects. This facilitates multiple-project resource scheduling.

In Part III, we will see how much of our project–control capability is based on this project model. Once approved, it becomes the ''static model'': the benchmark against which we measure what actually occurs.

6

Defining Contingency, Accuracy, and Risk of Cost Overruns

CONTINGENCY: AN IMPORTANT PART OF EVERY ESTIMATE

All those who prepare cost estimates share the common dread of being called upon to answer the question, "How much contingency should I add?" In actual practice, the answer to that question really depends on the answers to two more specific questions:

1. How much contingency is really needed?
2. How much contingency can we get management to accept?

The whole subject of contingency deals with uncertainty. It is therefore appropriate that most project and cost engineers experience a lot of uncertainty when dealing with it. We are usually uncertain about (among other things):

What contingency is really for
How to define it
How it relates to risk
How it relates to estimate accuracy
How to estimate it
How to control it

How to explain it
How to justify it
Where to put it
What to call it

Trying to explain contingency to management is a lot like a visit to the dentist: the best that can happen is that you come out the same way you went in—all other possibilities are increasingly painful. Even the word "contingency" has negative emotional connotations. It implies risk and uncertainty that cause fear.

Like it or not, contingency, risk, and estimate uncertainties are all very relevant today. Capital projects today are taking place in an environment of greatly increased financial risk, and of increased emphasis on budget levels, cost control, and effective use of management tools. Projects today, as compared to those in the past, are often:

Marginal, posing more risk to profitability
Longer term, with profit further in the future
Based on new technology, with associated risks
Subject to larger market fluctuations

Therefore even for small projects, there is a need to develop a clear and practical approach to handling contingency.

THE UNCERTAIN NATURE OF ESTIMATES

Although most project leaders and managers deal with estimates every day, it is worthwhile to reiterate just what an estimate really is. According to Webster's, to estimate is to "calculate approximately the worth, size, or cost." So an estimate is, by definition, an approximation; a prediction of what is going to happen. Therefore, *all estimates, by definition, contain a significant amount of uncertainty*. In spite of this indisputable fact, most project leaders and estimators tend to "sell" their estimating work by making the estimate look as credible as possible, often by downplaying (or even covering up) uncertainties. This is a natural tendency: it's hard to find a "how to succeed" book that recommends telling the boss how uncertain we are about the work we have just done, which is to be the basis for a major decision.

The reason for this paradox is found in the way estimates are used. The basic purpose of estimates is to aid in decision making. Company management uses estimates for decisions such as:

Selection of projects for further study
Selection of the preferred design approach
Economic analysis
Budget approval and commitment of funds for research, development, engineering, procurement, and construction
Selection of contractors and suppliers
Approval of major changes

Project managers and cost engineers also use estimates for the very important purpose of cost control.

All these estimate users require accuracy in their estimates. They want to feel confident that they have good estimates on which to base their difficult decisions. Therefore, estimate uncertainty, accuracy, risks of overrun, and contingency are subjects that are apt to be avoided. The project leader doesn't want to talk about it, and the manager doesn't want to hear about it. This would be a happy situation were it not for the unfortunate result that *many important investment decisions are made without proper recognition of the financial risks involved.*

It is, therefore, important that project leaders be able to quantify, control, and communicate risk, uncertainty, and contingency. Whether we like it or not, estimating means dealing with uncertainty and decisions based on estimates mean dealing with risk.

A DEFINITION OF CONTINGENCY

Before we can measure something, we have to be able to define it. Unfortunately, everyone knows what contingency is, but everyone thinks it is something different. For example, there is one school of thought (illustrated by Figure 6.1) that maintains that contingency is a source of funds, there if you need it, which shouldn't be used under ordinary circumstances. The problem with this view is that the contingency funds have a way of getting spent before the project is over.

Another view of contingency (illustrated in Figure 6.2), suggests that contingency should be rationed out to project managers or contractors when they show that they have done their jobs, and the requirement for extra funds is not their fault. This approach is intended to make them work for any releases of contingency. The problem here is apt to be that the contingency is needed anyway, but that the need for it may be concealed.

Another popular approach, illustrated by Figure 6.3, maintains that, "contingency is not required on our projects, and we don't allow it."

Contingency: Funds which are not
to be used under
ordinary circumstances

Figure 6.1   The "piggy bank" view of contingency. Contingency is, in this view, funds that are to be used only for emergencies.

Contingency:
Something you
can have if you've
been good

Figure 6.2   The "cookie jar" view of contingency. Here, contingency is something you can have if you can justify it.

Figure 6.3   The "ostrich" view of contingency. Here, the existence of contingency is completely denied.

Naturally, these projects do get by without contingency funds, although they often have what amounts to contingency funds given a different name.

At the other end of the spectrum, we can identify the approach in which contingency funds are expected to be spent, and are spent, freely. This approach, illustrated by Figure 6.4, naturally tends to make cost control a problem.

As a result of these diverse views, which sometimes co-exist in the same organization, many project leaders adopt defensive strategy shown in Figure 6.5. Just like the family dog, who buries bones where only he

Contingency : There to be spent

Figure 6.4   The "blank check" view of contingency. Here, contingency is seen as funds that are there to be spent.

Estimators "bury" contingency
where only they can find it

Figure 6.5   Estimating contingency: the "backyard bone-burying" approach. Estimators "bury" contingency where only they can find it.

can find them to assure that they are there when needed, we "bury" contingency funds in our estimates and cost forecasts. The result, of course, is hidden contingencies, over-estimated project costs, inflated budgets, and ineffective cost control.

So it is an important first step just to define contingency. The following definition of contingency (presented in Chapter 5) has proven to be clear, useful, and generally acceptable: *Contingency is a provision for those variations to the estimate basis that are likely to occur but that cannot be specifically identified at the time the estimate is prepared.*

There are three key elements to this definition:

1. *The basis of an estimate is the design, the execution plan, and the pricing levels* that are defined or assumed at the time the estimate is done. The actual costs will vary from the estimate to the exact extent that variations are experienced in the basis. Therefore, we can explain all uses of contingency funds in terms of variations to the estimate basis.
2. *Contingency covers those variations that are likely.* In other words, low probability occurrences, such as *forces majeures*, are not covered by contingency, even if they could have a major cost impact. Since the project budget should include contingency, it should only provide for those added costs which have a reasonable probability of occurring.
3. *Contingency covers those variations that cannot be specifically identified.* In other words, if we can identify a likely variation, such as a major design change, it belongs in the base estimate, not contingency.

Some examples of likely variations that would be covered by contingency are:

Design changes (excluding scope changes)
Variations in the way the project is managed
Estimating variations and errors

Equally important are the potential variations that are not covered, by contingency. These are the low probability, high cost-impact risks that are not likely to occur but, if they do, can cause enormous disruption. Examples of such risks are:

Scope changes
Major changes in schedule milestone dates
Unprecedented fluctuations in exchange rates, inflation rates, etc.

Although it would clearly be inappropriate for the project leader's budget to contain funds for such risks, it is entirely possible that top management might wish to know what the financial impact of such risks would be, and to have some funds in reserve for that purpose. Other investors in the project, or its financiers, might want to know, "How far could I possibly be sticking my neck out?" To answer this question, we define "manager's reserve": the extra funds required to produce a very high confidence that the project will not overrun.

To illustrate these concepts, we might plot cumulative probability against project cost (see Figure 6.6). A cumulative probability of 50%

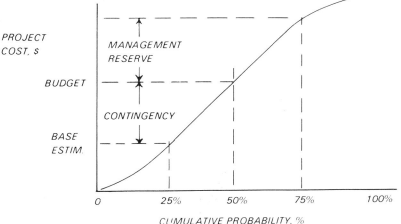

Figure 6.6 Project cost vs. probability. The estimate accuracy determines the slope of the curve.

means that there is a 50% probability that the final cost will be equal to or less than the cost shown on the curve at that point. If the project leader's budget is to be at the 50/50 point (i.e., to have an equal chance of overrunning or underrunning) the contingency will be the difference between the cost at the 50% probability point and the base estimate excluding contingency. Management reserve will be the difference between the budget and the cost at the point of desired confidence.

WHY CONTINGENCY IS NECESSARY

The above definition raises some interesting questions: "Why should contingency be required?" "Why aren't the base estimates at the 50/50 point?" "Why are the variations likely to produce a net increase in the project cost?" Or, to put it another way, "Why does Murphy's Law always seem to be proven right on my projects?" To answer these questions, we need to use the statistical concept of "skewness." If we were to plot the frequency with which we expect to experience each possible value of project cost, we would describe a "frequency distribution curve," as shown in Figure 6.7. We would find that there is one value that will occur more often than any other value. This "most likely" value is the highest point on the frequency distribution curve, and is usually our base estimate. Thus, if we look at a component of the estimate (such as material costs), at an estimating variable (such as labor cost per hour), or at the total project cost, we generally find that we have, using the "most likely" value, estimated the cost that is most likely to be correct if the

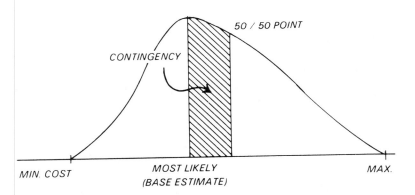

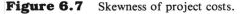

Figure 6.7   Skewness of project costs.

basis of the estimate remains unchanged. Unfortunately, variations to the estimate basis are almost certain to occur.

If we now consider the range of possible variations, we note that the best-case scenario gives us a minimum cost somewhat below the base estimate. There is, of course, a minimum cost level below which we cannot go. However, when we consider the maximum possible cost under the worst-case scenario, we find that there is apt to be almost no limit to how bad things can get and how high the cost can go. For example, although it is impossible for the project to be completed at 100% below budget, there are many projects that have ended up several hundred percent over budget.

So skewness is an inherent part of project-frequency distributions. Most project data is skewed to the right, meaning that there is a greater probability that our base estimate will overrun than that it will underrun. This is shown in Figure 6.7, in which we see that the 50% probability point is to the right of the base estimate. The difference between the most likely estimate and the 50/50 estimate is contingency.

THE ESTIMATOR'S DILEMMA

Having formulated an acceptable definition of contingency, the estimator is now faced with the task of quantifying contingency: that is, of defining how much it should be for a given estimate. However, if he or she accepts the definition provided above, and begins to define a value for contingency, the estimator is faced with a dilemma: How can one define that which, by definition, is undefined? In other words, if contingency is for those variations that cannot be defined, how are we supposed to define them sufficiently so that we can put a value on contingency?

This dilemma is confusing enough, but if we try to work with it, it gets even worse. It is apparent that some form of statistical analysis is necessary if contingency, risk, uncertainty, and estimate accuracy are to be quantified. That itself is a problem: for some reason, very few people are comfortable with applied statistical methods. For example, these people say that a statistician will insist that a man with one foot encased in a block of ice and the other planted on a bed of burning coals is, on the average, comfortable! Most company procedures, in addition, require that a cost estimate or budget have one value. Ranges of values are not permitted or appreciated.

On top of all this, it is also important to know that *the perfect method for analyzing risk, uncertainty, estimate accuracy and contingency does not exist!* Therefore, whatever is done to improve the way things are done will still be vulnerable to valid criticism. So what is the estimator to do?

DEFINITION OF RELEVANT TERMS

Improving the way we define, communicate, and manage contingency requires the same type of program we would apply to any other area of management. Key elements are as follows:

Obtain top management support and endorsement
Define and agree on all relevant terminology
Develop and use practical methods for quantification of risk, uncertainty, accuracy, and contingency
Develop and use practical methods for the control of risk and the administration of contingency funds
Monitor effectiveness of methods, and refine as necessary

Contingency Terms and Concepts

Let us start with the definition of terms.

Risk: The probability that a certain undesirable outcome will occur. Example: There is a 10% risk that the final cost will be more than 30% above the budget.

Uncertainty: The range of values within which the actual value is expected to fall. Example: The uncertainty in our estimate of wage rate is such that we expect it to fall between $10 and $40 per hour.

Variation: The difference between the design, planning, and cost basis of the estimate, and the actual design, execution, and cost of the project. All variations in cost can be explained by variations to the estimate basis. (Example: Design changes resulted in a 5% cost increase. Increased use of subcontract labor resulted in a 9% cost saving. Escalation was less than anticipated, resulting in a cost savings of 6%).

Estimate accuracy: The confidence limits within which there is a specific probability that the actual cost will fall (see Figure 6.8). (Example: The accuracy of this estimate is such that there is a 90% probability that the final cost will be within plus or minus 10% of the estimate.)

Note that this definition of estimate accuracy means that it is not adequate to describe accuracy as, "plus or minus 10%." One must always specify the probability that the actual cost will fall within the confidence limit.

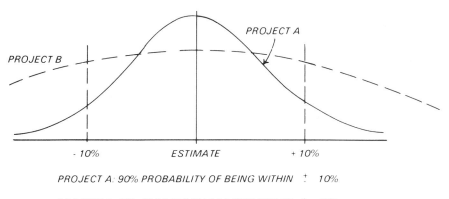

PROJECT A: 90% PROBABILITY OF BEING WITHIN ± 10%

PROJECT B: 60% PROBABILITY OF BEING WITHIN ± 10%

Figure 6.8 Defining estimate accuracy: the confidence limits within which there is a specified probability that the actual cost will fall.

Statistical Terms and Concepts

Since we will be using some simple statistical concepts in our work with contingency, it is also appropriate to review some basic statistical concepts and terminology.

Frequency distribution curve (*or probability density function*): This type of plot, shown in Figure 6.9, describes the frequency with which a given value will be experienced. The higher the curve at a given point, the more often we can expect that value to occur. The well-known "bell curve" is an example of a symmetrical frequency distribution.

Since the likelihood that a specific, discrete value of cost will actually be experienced is quite small, it is more useful to think in terms of the probability that the actual cost will be equal to or less than the given value. That probability is indicated by the area under the curve.

The frequency distribution curve can be described by several, very useful parameters (see Figure 6.9):

The *most likely* value is at the peak of the curve, and is the single value that will be experienced most often. It is also called the mode.

The *median* value is the midpoint of the distribution of all possible values. It is the "50/50 point," in that half of all possible values are above it, and half below.

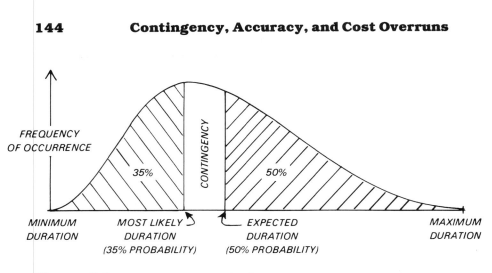

Figure 6.9   The frequency distribution curve.

The *mean* value is the average of all data points in the distribution.

The *maximum value* is that value for which there is a zero probability that any actual value will exceed it. It corresponds to the worst-case scenario.

The *minimum value* is that value for which there is a zero probability that any actual value will be less. It corresponds to the best-case scenario.

The *range* is the difference between the maximum and minimum values.

The *standard deviation* describes the *"closeness of fit"* of the distribution. If the standard deviation is large, the points are widely scattered. If the standard deviation is small, the points are clustered about the middle.

From a project leader's viewpoint, the most important parameters are the most likely, median, and standard deviation.

QUANTIFYING ESTIMATE RISK, UNCERTAINTY, AND CONTINGENCY

Basic Approach

There are only two ways to define the probability function that provides estimate accuracy, risk of overrun, and contingency requirements. The first way might be described as "looking back," that is, by the use of historical data from many past projects. The second way might be described as "looking forward" that is, by probabilistic analysis of all possible outcomes of the project at hand. These two approaches are described as follows.

Analysis of Historical Data

This approach is based on the assumption that the project at hand is essentially similar to the many past projects represented by the data. To state it a different way, it assumes that the past is the best predictor of the future. This assumption is likely to be valid when the project at hand is considered typical. In such cases, past data from actual projects is the best predictor of what can be expected from future projects.

To prepare the cost/probability curve using historical data, the data must be adjusted to a consistent basis, and then analyzed using standard calculations for mode, median, and standard deviation. When these parameters are known, the distribution is defined and one can estimate the probability of overrun. The distribution variable that is selected can be total cost, unit cost (e.g., $ per unit of production), or contingency amount (calculated as the difference between actual and estimated costs).

Successful application of historical data analysis requires that the project be similar to past projects, that the data used is reliable and appropriate as a predictor of future performance, and that no special risks need to be reflected in the analysis.

Probabilistic Cost Analysis

In probabilistic cost analysis, the project at hand is analyzed by the process of looking into the future and examining what possible outcomes exist. In other words, we synthesize the cost vs. probability function from the specific risks and uncertainties that are associated with the project. This approach is preferred in those situations in which, (1) the project at hand is not similar to many past projects, (2) in which the risks it will be facing are specific and atypical, and/or (3) in which the available data is not adequate.

There are a number of methods for probabilistic cost analysis. All have advantages and disadvantages; none are perfect. Some of the more widely used methods are described below. In these descriptions, we refer to the estimate excluding contingency as the "base estimate," and when contingency is added it becomes the "budget estimate."

The Monte Carlo Simulation

In the Monte Carlo Simulation (see Figure 6.10), the user defines equations that duplicate or model the way the estimate has been done. For example, the cost of excavation may have been estimated using this equation:

Cost of excavation labor = (cubic yards)(work-hours/cy)
(cost/work-hour)

The terms on the right side of the equation are "cost variables." A probability distribution is then defined for each cost variable, by specifying the maximum and minimum values that the variable could experience. A frequency distribution for the value of that variable can be deduced. This process is repeated until the entire cost estimate is represented by the cost model, and each cost variable has been assigned a maximum and minimum value.

The next step in the method is to simulate many possible outcomes of the project, using the cost model and the frequency distributions for each variable. This is done by selecting a random number (as one might spin a roulette wheel in Monte Carlo) that is used to select a value for each

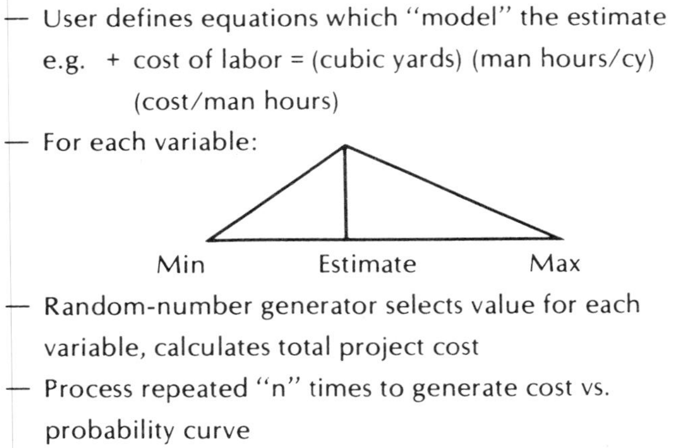

— User defines equations which "model" the estimate

e.g. + cost of labor = (cubic yards) (man hours/cy)

(cost/man hours)

— For each variable:

Min   Estimate   Max

— Random-number generator selects value for each

variable, calculates total project cost

— Process repeated "n" times to generate cost vs.

probability curve

Figure 6.10   The Monte Carlo simulation. In this simulation, the user defines equations that model the estimate. A random-number generator then selects values for each variable, and calculates the total project cost. This process is repeated perhaps 1000 times to generate the cost vs. probability curves.

cost variable, according to its frequency distribution. When a value has thereby been selected for each variable, a possible outcome of project cost is calculated. This is one datapoint on the frequency distribution curve for total project cost.

The process of selecting a random number, selecting a value for each cost variable, and calculating a possible outcome of project cost is repeated hundreds or even thousands of times, until a complete probability distribution of total project cost is generated. This distribution can then be used to calculate accuracy, risk of overrun, and contingency requirements. The Monte Carlo simulation is computerized, and there are a number of good programs available that are relatively easy to learn and use.

Hand Calculations Using Statistical Approximations

In those situations in which it is inconvenient or inappropriate to use a computer program, there are a number of statistical-analysis techniques that can be done manually. These can also give an approximate result that will prove useful.

PERT approximations   One of the best-known statistical approximations is provided by PERT (Program Evaluation and Review Technique; see Figure 6.11). Used to analyze cost or schedule, PERT gives us the following formula:

$$\text{Expected value} = \frac{\text{minimum} + 4(\text{most likely value}) + \text{maximum}}{6}$$

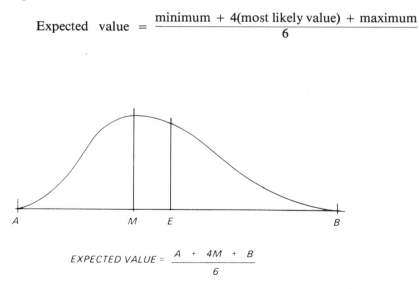

$$\textit{EXPECTED VALUE} = \frac{A + 4M + B}{6}$$

Figure 6.11   PERT approximations.

The expected value (i.e., median or 50/50 point) can be calculated for each cost variable or component of the estimate, and then the expected value of total project cost is simply the sum of the expected values for each cost component. Contingency is then the difference between the expected value of project cost and the base estimate. This method provides a cost and/or schedule contingency for each planned activity or cost category. It is easily computerized using a spreadsheet. In some cases, schedule or cost data can be imported from whatever program was used to prepare it, and then automatically analyzed in the spreadsheet, using the PERT technique.

Cost-loaded network models   Some projects are planned and controlled by the use of cost-loaded networks, in which the cost of each activity is identified, consistent with its duration and resource requirements (see Figure 6.12). These cost-loaded networks can be thought of as a form of project model. One way that schedule and cost contingency can be addressed in an integrated fashion is to define a model for which the duration and cost for each activity is at the 50% probability level. Schedule and cost contingency can then be defined as the difference between the end date and final cost predicted by the 50/50 model and the original, target model.

Approximating the standard deviation   The standard deviation is important, as it is the parameter that tells us the shape of the cost vs. probability curve and allows us to determine the probability of any cost level being overrun (see Figure 6.13). We can approximate the standard deviation

— Cost assigned to each activity

— Network and estimate for cost-time "model"

— Cost and duration for each activity can be set at

   "most likely" or at "50/50" level

Figure 6.12   Cost-loaded networks, in which a cost is assigned to each activity, a network and estimate for cost-time model is developed, and cost and duration for each activity can be set at "most likely" or at "50/50" level.

Standard deviation (σ) describes "spread" of probability curve

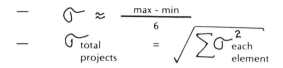

$$-\quad \sigma \approx \frac{max - min}{6}$$

$$-\quad \sigma_{\substack{total \\ projects}} = \sqrt{\sum \sigma_{\substack{each \\ element}}^{2}}$$

Figure 6.13   Useful approximations of standard deviation. The standard deviation (σ) describes the "spread" of probability curves.

with that formula. If we use this formula to calculate the standard deviation for each component of the estimate, and if the total project cost is the sum of the component costs, we can calculate the standard deviation of the distribution of all possible project costs by using the following formula:

Variance = (Standard Deviation)²

Variance (total cost) = $\sum$ Variance (components)

Decision-tree analysis   A "decision tree" displays the alternative courses of action or outcomes that are possible at a certain point in the project. If a probability and a cost impact is attached to each outcome, the most probable result can be identified. Decision trees are a good tool for analyzing specific risks, though they do not allow us to analyze the combined effect of many different variations.

Expected-value analysis   This type of analysis is similar to the decision tree in that specific risks and cost outcomes are identified. A probability is attached to each one, and the "expected value" is calculated as the product of the probability and the cost outcome. In some applications, these expected values are summed, to get the expected value of the total project.

Utility theory   All of use have a certain bias toward risk. We may have a high willingness to accept risk (the gambler mentality), or an aversion to risk. Since all capital projects involve some consideration of risk and return, utility theory may be used to calculate the risk/reward profile of the corporation and key managers. This can then be used as a guideline in evaluating projects. Utility theory provides a supplement to the other risk analysis methods.

Pitfalls to Avoid

One common error in dealing with probabilities involves the basic concept of combined events. The probability that events A *and* B will occur is the product of the probabilities of each event occurring, that is:

$$P(A \text{ and } B) = P(A) \times P(B)$$

If we wish to describe the probability that A *or* B will occur, that probability is the sum of the probabilities of each event, that is:

$$P(A \text{ or } B) = P(A) + P(B)$$

Since some project risks are considered to act independently, and others usually act in concert with other risks, it is important to consider when probabilities or expected values should be added or multiplied.

The Problem of Independence

The Monte Carlo and PERT techniques described above require that the individual cost variables or cost components be independent of each other, that is, that variation in one variable be totally unrelated to variations in others. While these methods give satisfactory results, it is, in fact, unlikely that the requirement for independence between variables will be met. It is therefore important, when designing cost models for use in probabilistic analysis, to try to define cost variables or components such that they are as independent as possible.

It is also possible to develop equations for calculating the most likely, median, and standard deviation for total project cost, in which the "covariance" between variables and components is reflected. The covariance can be thought of as a correlation coefficient between the variables. Guidance on methods for doing this can be found in texts on statistics and correlation theory.

ADMINISTRATION OF CONTINGENCY

Once a project is approved and work begins, cost control becomes the prime consideration. Contingency, being a major component of the budget and control estimate, must be administered and controlled like any other cost account.

Effective cost control requires that a cost forecast be made that reflects the outcome that is foreseen at that time. The contingency that is forecast to be required to complete the project should reflect the uncertainties existing at that time, just as we saw in preparation of the budget

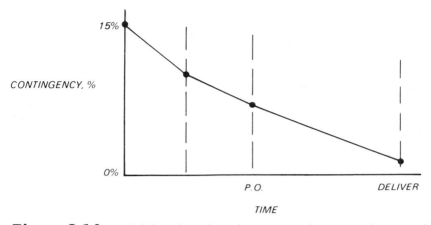

Figure 6.14 Administration of contingency (contingency = forecast of remaining uncertainty).

estimate. Therefore, reductions in contingency can be planned. Major reductions in contingency requirements can be made when milestones are reached, with the resulting major reductions in uncertainty.

Project contingency can be allocated to major cost codes, according to the uncertainty associated with each one. For each cost code, a schedule is prepared, showing the major milestones in uncertainty reduction as illustrated in Figure 6.14. For example, placement of major purchase orders results in considerably greater confidence in the forecast of escalation. At each milestone, we can define the contingency that will be required to cover the remaining uncertainty. Between these milestones, we draw a straight line. We now have a contingency rundown plan that can be used as a guide when setting contingency for the current model, as well as a basis for monitoring contingency funds.

A key point in contingency administration is that *the contingency required to complete the project should always be determined by the risks and uncertainties that lie ahead*, not by consideration of how much contingency has been "used" to date and how much therefore is "left."

CHAPTER SUMMARY

Uncertainty and risks are as much a part of project work as time and material. Contingency is the way we deal with uncertainty and risk, and it should be addressed in the same way we deal with any other project-management problem: firmly, clearly, and with the best technology available. Although the available methods are not perfect, they can go a long way toward improving current practices.

Because there are, in general, negative emotions associated with uncertainty and risk, project leaders should take the initiative in establishing the terminology and methods required to assure that investment and project-management decisions are made with proper recognition of the financial risks involved.

III

MANAGING MULTIPLE SMALL PROJECTS

7

Multiple Project Resource and Task Scheduling

INTRODUCTION TO MULTIPLE-PROJECT PLANNING

We've all seen the problem of the multiple project-resource manager. Imagine a group of twenty-five highly expert consultants. Each consultant works with three or four other consultants on a project. So when considered alone, the resource requirement of each project is well within the total availability of the group. But what do we do when the group's total workload usually consists of ten projects? The problem is that the total demand of ten active projects may well exceed the availability of one or more resources.

Who keeps track of the total demands on each resource of all the multiple projects? Who makes sure that the completion dates promised for each project are consistent with the availability of resources? Who sets priorities and juggles schedules when multiple projects are competing for the same pool of resources? Indeed, many companies find these problems to be very difficult.

Parts I and II of this book have dealt with the techniques for planning, activity scheduling, resource scheduling, and cost estimating for a single small project. Multiple project scheduling and analysis cannot take place unless each small project has been planned in a consistent and complete manner. However, once that has been done, the ingredients are in place for the powerful process of multiple-project planning and scheduling.

GOAL: CREATING THE MULTIPLE-PROJECT ACTIVITY AND RESOURCE SCHEDULE

This procedure is directed to the multiple project manager whose view is of numerous projects being performed by numerous resources. What this manager needs is a single schedule showing all of the projects under her control. In effect, this manager's "project" is to get all her projects done— on schedule, and with available resources if possible.

A multiple-project activity and resource schedule must:

Schedule tasks realistically so the requirement for resources does not exceed availability.
Provide early warning of the need to reschedule deadlines, change workscopes, increase resources, or work overtime.
Provide specific information on the current and planned workload for each resource, including all currently scheduled projects.
Be responsive to constant changes in workload, priorities and resource availability.
Be consistent with project priorities.

When we consider the problem of preparing the multiple small-project schedule, we can see that the principles of CPM, although essential for planning each individual project, have less relevance to the multiple-project plan.

To illustrate this point, consider a facilities management group serving a research laboratory. Each task is represented by a "work order" that typically takes about four days, and each work order is independent of the others. Each month, about 100 of these work orders are completed. A network plan for a project of one month's work would therefore look like Figure 7.1 where all activities (i.e., work orders) are independent and there are 100 parallel paths through the plan.

We might ask what role the critical path through this network can play. The critical path in this case is four days long—if we had unlimited resources (which a plan should always assume) the entire month's work would be completed in four days.

MONTHLY MAINTENANCE PROGRAM

CPM-PLAN FOR 100 WORK ORDERS

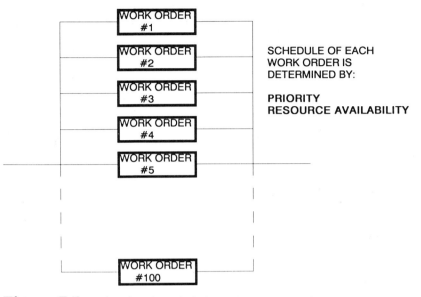

Figure 7.1   The plan for 100 independent small projects consists of 100 parallel paths. Therefore, the critical path is the length of the longest single activity and is not meaningful.

Of course, there are two important variables we are neglecting here:

Activities (i.e., work orders) have different priorities that will determine the order they get done.
Resources are not unlimited.

If we can apply these variables effectively to our multiple-project, multiple-path plan, we can produce a schedule that will show when each activity (i.e., work order) can be done, consistent with limited resource availability and our priorities.

METHODS FOR CREATING A
MULTIPLE PROJECT SCHEDULE

There are three types of situations in which a multiple project schedule must be prepared, and there is a method to handle each one.

1.) *When a project manager wishes to combine dependent subprojects to create a detailed plan for the whole project.* In this case, each subproject may have its own pool of resources. Here, "level two" plans (i.e., project management level of detail) are combined to create an overall project plan also at level two.

2.) *When the multiple project manager wishes to combine independent subprojects in order to create a detailed multiple-project plan.* In this case, each project uses the same pool of resources. Here, "level two" plans are combined to create a multiple-project plan also at level two.

3.) *When subprojects must be summarized to create a management-level multiple project master plan.* In this case, each project may or may not use the same resources. Here, "level two" plans are combined to create a multiple-project plan at "level one," the management summary level.

Each of these situations, and the method to handle it, are discussed below.

Combining Dependent Subprojects to Create an Overall Project Plan

Consider the case of a project leader who must coordinate the work of several functional groups in order to get the work for a project done. A project is actually a group of subprojects from different functional groups—where each group depends to some extent on work done by the others. One way to handle this problem is to have each functional group create a plan (i.e., subproject) for its own work.

For example, imagine an engineer planning a project in which a considerable amount of design-engineering work must be done. Using a personal computer with simple "low-end" software (see Chapter 15) each of the functional groups prepares a plan for its own work (see Figure 7.2a). Our project engineer receives a copy of the data file for each of these functional group plans, which can be considered to be a subproject of the total project.

Let us assume that the functional groups involved in the project are Process Engineering, Mechanical Engineering and Electrical Engineering. Each group has its own pool of resources. However, one or more of the

FUNCTIONAL GROUP "A" - PROCESS ENGINEERING

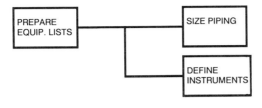

FUNCTIONAL GROUP "B" - MECHANICAL ENGINEERING

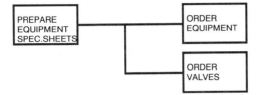

FUNCTIONAL GROUP "C" - ELECTRICAL ENGINEERING

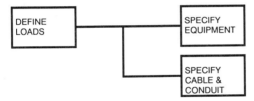

Figure 7.2a   As a first step in multiple project scheduling, individual sub-projects are prepared by each functional group to describe the work by that group.

activities in each subproject are dependent on activities in another subproj-ect. For example (see Figure 7.2b) the results of the first process-engineering activity (e.g., equipment lists) are required for the first mechanical engin-eering activity to begin. Likewise, the results of the first mechanical en-gineering activity (e.g., equipment specification sheets) are required for the first electrical engineering activity (e.g., calculate power loads).

Using the technique of combining subprojects, a master plan for the entire project can thereby be created. The technique usually involves:

Defining each subproject separately
Loading the first subproject

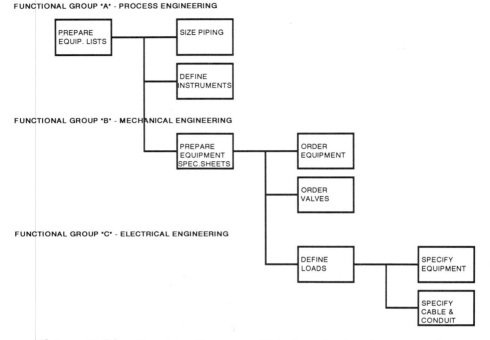

Figure 7.2b   Constraints between activities in each subprojects are established to create the multiple-project plan.

Combining the second subproject
Defining the constraints between activities in the first and second subprojects
Combining the third subproject
Defining the constraints between the first, second and third subprojects

Once the combining is complete, resource leveling can be performed in the usual way, defining a schedule that can be achieved with available resources.

Combining Independent Subprojects

Consider the case of a multiple small-project manager who must coordinate the work of one functional group as they work on a number of unrelated projects. One way to handle this problem is to create a plan for each small project, using the same pool of resources for all projects.

For example, imagine a multiple-project supervisor planning the maintenance work to be done over the next month in a power plant. Using a personal computer with simple "low-end" software (see Chapter 15) the maintenance supervisors each prepare a simple plan for each maintenance project. Note that each plan is prepared using the same pool of shared resources. Our supervisor receives a copy of the data file for each of these maintenance project plans. Each of these plans can be considered to be a subproject of the total maintenance program for the month.

Using the technique of combining subprojects, a master plan for the entire month's maintenance can be created. The technique usually involves:

Defining each subproject separately, using the resource pool
Loading the first subproject
Combining the second subproject, identifying that it shares the same resource pool
Combining the third subproject, identifying that it shares the same resource pool

Once the combining is complete, activities (i.e., maintenance tasks) can be prioritized. Resource leveling can now be performed in the usual way, defining a multiple-project schedule that can be achieved with available resources.

Summarizing Subprojects

Consider the case of a manager of a functional group that performs unrelated multiple small-projects. As before, one way to handle this problem is to create a plan for each small project, using the same pool of resources for all projects. The difference between this case and the previous two cases is that, in this case, subproject data is summarized into a single activity on the multi-project master plan.

For example, imagine the manager of a group of computer programmers. This manager must know:

The total workload of the group
When spare capacity is available
What projects are active
What projects are planned
The scheduled start date, finish date, duration, resource requirements and cost for each active and planned project

Using a personal computer with simple "low-end" software (see Chapter 15) the lead programmer on each small project prepares a simple plan for each project (see Figure 7.3a). Note that each plan is prepared using the same pool of shared resources. Each of these plans can be considered to be a subproject of the group's total workload for the month.

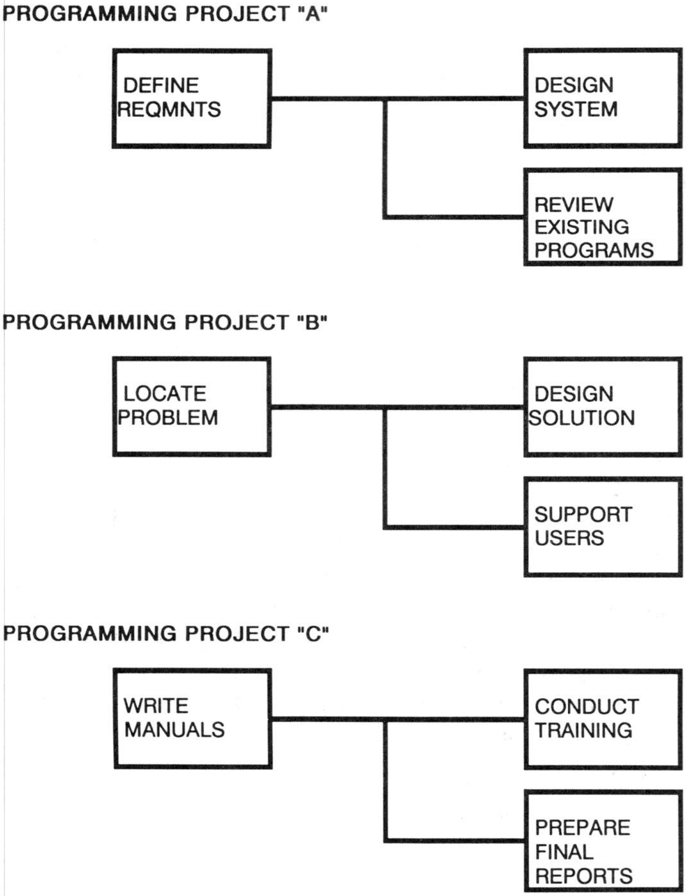

PROGRAMMING PROJECT "A"

DEFINE REQMNTS — DESIGN SYSTEM — REVIEW EXISTING PROGRAMS

PROGRAMMING PROJECT "B"

LOCATE PROBLEM — DESIGN SOLUTION — SUPPORT USERS

PROGRAMMING PROJECT "C"

WRITE MANUALS — CONDUCT TRAINING — PREPARE FINAL REPORTS

Figure 7.3a   Each programmer creates a plan for the small project on which he or she is working.

PROGRAMMING PROJECT "A"

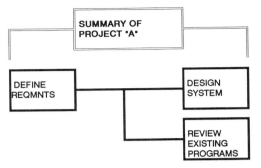

PROGRAMMING PROJECT "B"

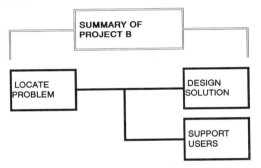

PROGRAMMING PROJECT "C"

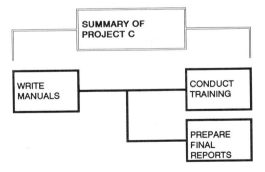

Figure 7.3b   Each small project is summarized with a single activity in the master, multiple-project plan.

Using the technique of summarizing subprojects, a master plan for the entire month's work can be created. The technique (see Fig. 7.3b) usually involves:

Defining each subproject separately, using the resource pool
Summarizing the first subproject into a single activity on the multiple-project plan
Summarizing the second subproject into a single activity on the multiple-project plan
Summarizing the third subproject into a single activity on the multiple-project plan

Once the summarizing is complete, activities (i.e., programming projects) can be prioritized. Resource leveling can now be performed in the usual way, defining a multiple-project schedule that can be achieved with available resources.

METHODS FOR RESOURCE LEVELING ACROSS MULTIPLE PROJECTS

Once a total project workload has been defined by combining or summarizing projects, an overall project, task and resource schedule can be prepared. Elements of a total workload analysis of a multiple project plan include:

Resource schedules showing the assignments of each resource on each task on each project
Total workload schedule showing how the total requirement for each resource compares with the availability
Indication of the schedule slippage (if any) caused by resource availability being less than required

The steps for multiple-project resource scheduling are similar to those for single-project resource scheduling—they are:

1) Resources are allocated to each activity on each project
2) Projects are combined to form a multiple project plan
3) Resource requirements are summarized in a histogram for each resource—showing the schedule of resource requirements over all the projects
4) Resource availability is defined as the total number of resources in the "resource pool"

5) Automatic resource leveling is performed in order to allow the program to calculate the schedule that can be achieved with available resources. Software features that are helpful include:

> *Priority-driven leveling*—the priority of the task or project determines the order in which they receive resources
>
> *Assignment scheduling*—the schedule for each specific resource working on each specific task is defined
>
> *Selective resource leveling*—specific resources can be included or excluded from the leveling calculation
>
> *Leveling within float only*—the critical activities (and therefore project completion) are not delayed as a result of leveling. Of course, this method may allow resource overloads to persist
>
> *Choice of leveling methods*—we can select the relative importance of such variables as priority, start date, and task ID in assigning resources
>
> *Complex resource availability*—the availability of a given resource can be varied with time
>
> *Complex resource requirement*—the requirement for a given resource can be varied with time
>
> *Time-limited leveling*—the leveling calculation does not delay the project but instead tries to reduce overload peaks
>
> *Automatic feedback*—of the results of multiple-project resource leveling to the schedules of the individual projects

(See Chapter 15 for a further description of these and other software features which are relevant to this calculation.)

6) The manager interacts with the multiple-project schedule using "Do It Yourself" techniques to find a more optimum schedule of resources, tasks and assignments. This interaction could include:

> Juggling project or task priorities and re-leveling
>
> Using "soft logic" to change constraints in order to move activities into periods of spare capacity
>
> Using float to increase activity durations—thereby reducing resource requirements while maintaining the same level of effort
>
> Changing the calendar (i.e., working overtime, weekends, shifts, holidays) in order to effectively increase resource availability
>
> Changing the resource assignments by substituting an underutilized resource for an overutilized one

Increasing the availability of overloaded resources

Defining arbitrary constraints between projects (e.g., we decide that project 101 will be completed before project 305 begins)

Defining arbitrary start and finish dates for projects

7) The multiple-project activity and resource schedule is finalized and agreed.

8) The details of the assignments of resources to tasks is fed back to the appropriate people (i.e., the resources who are to do the work and the people for whom the work is being done).

9) The individual project models are kept up-to-date by those responsible, and updated information is used to keep the multiple-project schedule up-to-date.

10) The multiple-project schedule is updated for changing priorities and workloads, changes to resource availability, completed projects, and new projects.

CHAPTER SUMMARY

In this chapter we explored how the single-project planning, scheduling and resource scheduling techniques learned in previous chapters can be applied to the multiple-project situation. By combining projects in various ways, a multiple-project plan can be created, representing the total workload of a group or large project. This can then be leveled in automatic and interactive ways, until an acceptable compromise between deadlines and resource limitations is reached.

8

Assignment Modeling: A New Method for Multiple-Project Resource Management

ASSIGNMENT SCHEDULING—COMPLEMENTS TASK SCHEDULING

So far we have discussed the application of project-management techniques to the management of the single small project. By treating each small project in a consistent way, it is possible to combine, summarize and analyze multiple-project data, as was discussed in the last chapter. However, there is another aspect of multiple-project management that we have not discussed, and that is the problem of resource management.

In most organizations we can identify two types of management function that contribute to the management of projects. These are:

Project Management: in which one is concerned with scheduling tasks. CPM is useful for those who have this perspective on a project.

Resource Management: in which one is concerned with scheduling the assignment of resources to tasks. The resource manager often provides the services that a project needs. Examples of this situation are found in the functional and matrix forms of organization where a project manager must rely on functional resource managers to provide the support they need to get projects done.

This chapter will discuss the technique of Assignment Modeling, a method for assigning resources to tasks in an optimum way.

The Project Management Perspective

The Critical Path Method (CPM) views the way people work as a series of dependent tasks that describe a project from start to finish. This view matches the view of the project manager, who must schedule and control the performance of tasks. The network diagram (i.e., PERT chart), bar-chart (i.e., Gantt chart), and the critical-path analysis are all useful and well-proven for task-scheduling and control.

On projects managed with CPM techniques, resources are usually a category of person or equipment. For example, a human resource might be "welder" and an equipment resource might be "crane." These resources are usually dedicated to the project and provided in quantity.

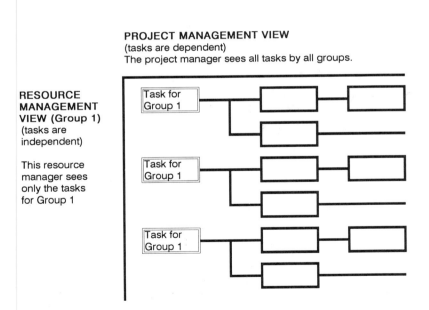

PROJECT MANAGEMENT VIEW
(tasks are dependent)
The project manager sees all tasks by all groups.

RESOURCE MANAGEMENT VIEW (Group 1)
(tasks are independent)

This resource manager sees only the tasks for Group 1

Task for Group 1

Task for Group 1

Task for Group 1

Figure 8.1   The project manager sees a project as a series of dependent tasks, going from start to finish. CPM-based project management software is appropriate for this view. The resource manager sees only a piece of each project, comprising many independent tasks. Assignment Modeling software is appropriate for this view.

The project manager needs to know that he needs 20 welders next week, he does not usually need to know their names.

The resource histograms that project-management software produces are useful in this environment, as are the calculations of resource-leveled schedules. Although CPM methods may be useful for the project manager, most resource managers agree that these methods are not effective tools for them.

The Resource Management Perspective

Things are different for the resource manager. Unlike the CPM model, resource managers view their workloads as a set of independent, largely unrelated tasks. Each of these tasks represents only a part of the project to which it belongs, and each belongs to a different project. To the resource manager, resources are individual people, such as Joe Smith, or equipment items, such as crane no. 3 (see Fig. 8.1).

RESOURCE SCHEDULING: KEY TO PRODUCTIVITY

A "resource" is a person, or equipment item that expends effort to do work. Resource scheduling is defining:

What specific resource
Should work on which specific task
Using which skill
At what level of effort
Working what percent of the time
Between which dates

Most projects require work by several resources. It is not until all those resources are ready to work that the task can proceed. And the project will not be complete until all the resources have completed the work required of them. Traditional methods for scheduling tasks are, therefore, of little value unless we back them up with effective methods for scheduling resources.

Resource scheduling is not only the key to successful projects, it is also a key to successful companies. Good resource scheduling is the basis for maximizing the productivity of people and equipment while minimizing their cost.

Few companies can remain profitable in today's highly competitive business environment without effectively managing the cost of resources.

Many have therefore attempted to cut costs and improve profitability by "downsizing": flattening the organization chart while reducing (sometimes drastically) the number of effective resources. Others have been remaining competitive by rapid growth, adding people or equipment to keep pace with demand. Both these scenarios require careful attention to the quantity and qualifications of the resource pool, and to how we assign them. These are problems that, with traditional methods, are difficult to address.

What does it take to have productive resources?

An organization containing the correct quantity of resources with the right mix of skills

The ability to schedule the assignments of resources to tasks in an optimum way

People committed to their schedule and performance goals

Good supervision of the work

These needs would be greatly helped by a resource scheduling method that could find the optimum way to schedule resources and also provide schedule and utilization performance data. Such a method would help assure that we have the right mix of resources and skills, that we have assigned resources to tasks in an optimum way, and have set realistic goals for each task.

Such a method exists. It is Assignment Modeling™. The following paragraphs will describe how Assignment Modeling was developed and how Artificial Intelligence made possible an elegant solution to the classic, vexing problem of resource scheduling.

RESOURCE SCHEDULING—KEY TO MANAGING ANY PROJECT

In a typical matrix organization, the small project manager must go to various functional groups to get work done on his or her project. Each functional group is, of course, headed up by a different resource manager who must respond to the needs of many projects.

The ability of small project managers working in matrix organizations to complete their projects on time depends heavily on the performance of resource managers. The required work will be done on schedule only if resource managers provide the required resources in the required quantity at the required time. It's a tough job scheduling resources. And until now, resource managers have not had much help.

RESOURCE SCHEDULING:
THE RESOURCE MANAGER'S VIEW

To illustrate resource management, let's consider a group of people working for an automobile manufacturer. This group does design work involving the car's brakes, steering and suspension.

The group's supervisor is a resource manager named Brenda Brown. Brenda must respond to the needs of many project managers, who care about their project and little about Brenda's workload or obligations to others. Brenda's situation changes constantly as new projects begin, existing projects grow in scope, priorities change, and people come and go.

Like all resource managers, Brenda must provide project managers with the maximum amount of quality work in the minimum amount of time. She also must provide the people in her group with appropriate assignnments, realistic performance objectives, and opportunities to learn and grow. Finally, she is responsible to company management to get the maximum work from the minimum number of resources—all this at a time of heightened concerns about employee satisfaction.

Resources Are Complex

A resource in this group is a design engineer named Jim Green. Like the others in the group, Jim has more than one skill. Jim is the group's best person for suspension design, but he also can work on brake design. In addition, he is in training to use the group's Computer Aided Design (CAD) workstation. When considering Jim for an assignment, Brenda must consider his experience at the type of work to be done. For example, Jim will do suspension work faster than anyone. Although he is qualified to use the CAD workstation, his inexperience at this type of work will cause him to take longer than someone else might.

In addition, Brenda must consider the vacation schedule and other periods, such as training and hospitalization, when resources are unavailable. Before assigning a resource to a task, Brenda also considers its existing workload. Should Jim be taken off an existing assignment to work on a new, high-priority task? If so, can someone else fill in for him, or should he go back to the original task after the new one is complete? If someone fills in on Jim's existing assignments, who fills in for the people who are filling in for Jim?

When we consider all the actual variables, the resource scheduling problem quickly becomes incredibly complicated. One can understand why many resource managers have concluded that there is no method or software that can address it effectively.

AI GIVES THE SOLUTION: ASSIGNMENT MODELING

Assignment Modeling—An Expert System

Although some have criticized Artificial Intelligence (AI) for failing to live up to expectations, it provides an elegant solution to the vexing problems of resource scheduling. The solution is a method called Assignment Modeling, and the AI application it uses is the technique known as Expert Systems.

Expert Systems attempt to capture the thinking process that an "expert" uses to solve a particular problem or reach a conclusion. Many successful applications use the idea of "expert" to mean simply a person who knows how to do a particular thing well. We can capture the expertise as a set of rules that form a "knowledge base." When programmed, the knowledge base becomes an "inference engine" that can apply the rules to any situation that is input.

A resource manager's job is to define an optimum "assignment schedule" for his group, that is to define:

What resource
Will work on which task
Using which skill
At what level of effort
What percent of the time
Between which dates

A resource manager uses rules to do this, and these rules can be programmed. For example, will a resource manager remove a resource from an incomplete existing task to work on a new one? A rule might say that the answer is Yes, but only under certain conditions:

If the new task is more important than the existing one
Or if the existing task has more time to spare than the new task
And if the required resource will be permitted to stop work on the existing task
Or if the required resource can be replaced on the existing task by another resource that can still finish it on time
And if the required resource is not on leave when the new task must be done

This type of rule-based algorithm makes possible the technique of Assignment Modeling: the interaction of a resource manager with an expert system to produce the optimum assignment schedule.

ASSIGNMENT MODELING—CALCULATING THE OPTIMUM ASSIGNMENT SCHEDULE

The Assignment Modeling Expert System— Simulates a Resource Manager's Thinking

The Assignment Modeling expert system thinks just as a resource manager does. Every resource manager knows that the only way to cope with the demanding, multi-project environment is to set priorities. The resource manager must decide what is the most important thing that needs to be done, and then decide which resource should do it. Once the resource manager takes care of the most important assignment, he considers the next most important assignment, and so proceeds until all requirements have been met.

The Assignment Modeling expert system works the same way. It ranks task requirements in order of importance, and resources in order of desirability to do each task. This process allows development of an optimum assignment schedule. The way in which a resource manager uses this calculation is simple.

Assignment Modeling: A Simple Process

The Assignment Modeling method is a process by which:

The resource manager inputs preference, resource and task information
The expert system proposes an optimum assignment schedule, based on
    the resource manager's input
The resource manager interacts with the expert system until an optimum
    assignment schedule is defined

The Assignment Modeling method uses software that runs on an ordinary IBM PC or compatible. The procedure described below, is used by Sagacity[tm] software, from Erudite Corporation.

The flow diagram, Fig. 8.2, shows each step in the Assignment Modeling method. These steps are described below.

Step 1: Define Preferences

Task Preferences   What makes a task important? Different people may have different answers. Some resource managers place the greatest importance on a task's priority. For example, a maintenance supervisor at a nuclear power plant knows that failure to attend to a "priority one" task could mean a serious accident. A manager scheduling manufacturing

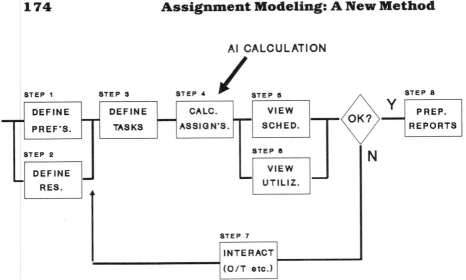

Figure 8.2 Assignment Modeling is an 8-step process, based on the inter-action of the expert system with the resource manager, to find an optimum assignment schedule.

equipment may consider all tasks to be equal in priority. To him, the most important task is the one that has the least time to spare.

The Assignment Modeling method allows the expert system to be adjusted for these preferences. It does so by allowing the resource manager to input task ranking preferences. The program asks: Would you prefer that the first task to be assigned is the one with the highest priority? . . . the one with the least time to spare? . . . the one that requires the scarcest skill?

The resource manager can then rank these selections in the appropriate order. For example, by accepting the order shown above, our example manager, Brenda Brown, shows she wants the highest ranking to be given to top-priority task assignments. If she has more than one top-priority assignment, then she wants the highest ranking to be given to the one with the least time to spare. If there is more than one such assignment, she wants the highest ranking to be given to the one that uses the scarcest skill.

Resource Preferences Once we have ranked all task requirements, the resources available and qualified to do each task must be ranked in order of desirability.

What makes a resource desirable? Again, the answer may not always be the same. A schedule-driven manager may prefer to assign the resource that can finish the task in the least time. Another supervisor, working on lump-sum contracts, may be more concerned about minimizing the cost of the work. He would prefer to assign the lower-cost resource, even if that resource takes longer. A manager whose company is "downsizing" may prefer to make assignments to concentrate the work. One whose company is growing may prefer to spread the work around. The Assignment Modeling method recognizes these differences through resource preferences.

The program asks: When more than one resource can provide the effort required by a task, do you prefer the resource that will complete the work in the minimum time? . . . for the minimum cost? . . . with busy resources (thereby concentrating the work) or with idle resources (thereby spreading the work)? . . . if two or more resources are still equally desirable, do you want to "break the tie" or have the program do it?

In this example, Brenda has stated that the resource that can do the work in the least time is the most desirable. If more than one qualifies, then she prefers the one that can do the work at the least cost. If more than one resource still qualifies, then she prefers the one that will result in concentrating the work in the fewest resources. Finally, if there is still more than one qualified resource, she prefers that the program make the decision and "break the tie."

Once set, resource and task preferences are unlikely to change from day to day.

Step 2: Define Resources

We define a resource in the following terms:

Name (name of the person or equipment item)
Title
Code (used for utilization analysis)
Availability period (date the resource comes into or will leave the pool)
National Holiday Set (which holidays does this resource take?)
Cost Per Hour (used for decision-making)
Skills (a resource can have several skills)
Standard Hour Multiplier for each skill (the time, from relative to average, that the resource requires when using this skill)

Note that the capability to provide multiple skills to a resource sets the assignment modeling method apart from other methods. Other methods generally assume that a resource has only one skill, and that all resources with that skill perform at the same level of effectiveness. In Jim Green's case, his skills would be Suspension Design, Brake Design, and CAD Operation, and he would have an appropriate Standard Hour Multiplier for each.

A resource also may be defined in terms of its "leave." A person may take leave as vacation, sick leave, training and so on. Equipment "leaves" may be downtime for maintenance or overhaul. The Assignment Modeling method allows leaves to be defined for any resource, and will advise the resource manager when a requested leave conflicts with existing assignments. If he grants the leave, the method will reassign that resource's work as necessary.

Step 3: Define Tasks

We describe a task in the following terms:

Task Name
Task Code (used for utilization analysis)
Task Priority (as defined by the resource manager)
Earliest Start Date (earliest date it could start)
Due Date

The work required by the task will be defined in terms of the workhours of effort required from each skill. For example, Brenda Brown's group may have a new task that requires:

900 workhours of Brake Design
150 workhours of Suspension Design
200 workhours of CAD Operation

Brenda also can specify what percentage of the time a resource will work on the task using each of these skills. She also knows that 600 hours of Brake Design must be completed before Suspension Design and CAD Operations can begin. The Assignment Modeling method allows this sequence of skills to be input as well.

Finally, Brenda can advise the Expert System how much flexibility it will have in finding the optimum way to make assignments. She can tell it that it is (or is not) permissible to split the assignment between two or more resources, and whether or not the work can be interrupted. She also can specify the resource that she wants to assign to the work (although to do so would prohibit the Expert System from finding out which resource is best).

The difference between Assignment Modeling and other methods is interesting. Unlike CPM, which describes a task in terms of its duration, Assignment Modeling views a task as a series of effort requirements by various skills.

Step 4: The Expert System Calculates the
Proposed Optimum Assignment Schedule

When any resource or task data changes in any way, the Assignment Modeling method identifies the need to prepare a new, optimum assignment schedule. Once the resource manager is ready to initiate the assignment scheduling process, it proceeds in four sub-steps:

1. Resources who conceivably could contribute to an optimum solution are "unassigned" from their existing assignments, thus making them available for reassignment
2. The program presents the proposed unassignments to the resource manager who may, if necessary, "lock" a resource to prevent reassignment
3. Once the resource manager gives the OK, the Expert System proceeds to rank tasks (and their individual skill requirements) and resources, and to make assignments accordingly. The Expert System is always trying to make assignments such that each task is completed before its due date. If necessary, it may "backtrack" to try splitting or interrupting assignments to avoid having an assignment finish after the due date
4. The program presents the proposed new assignment schedule to the resource manager who may accept it, or make further changes as described in Step 7

Step 5: View and Analyze the Proposed Assignment Schedule

The Assignment Modeling method presents the assignment schedule with two views: a Resource View (Fig. 8.3) and a Task View (Fig. 8.4). Each resource or task also can be viewed in detail, as shown by Figs. 8.5 and 8.6. These views show how well the resource manager is doing at meeting deadlines.

Step 6: View and Analyze Resource Utilization

The Assignment Modeling method also presents analyses of how well the resource manager is doing at using resources effectively. The program provides histograms showing total resource utilization as well as utilization of each resource, skill, and so on (see Figures 8.7, 8.8, 8.9).

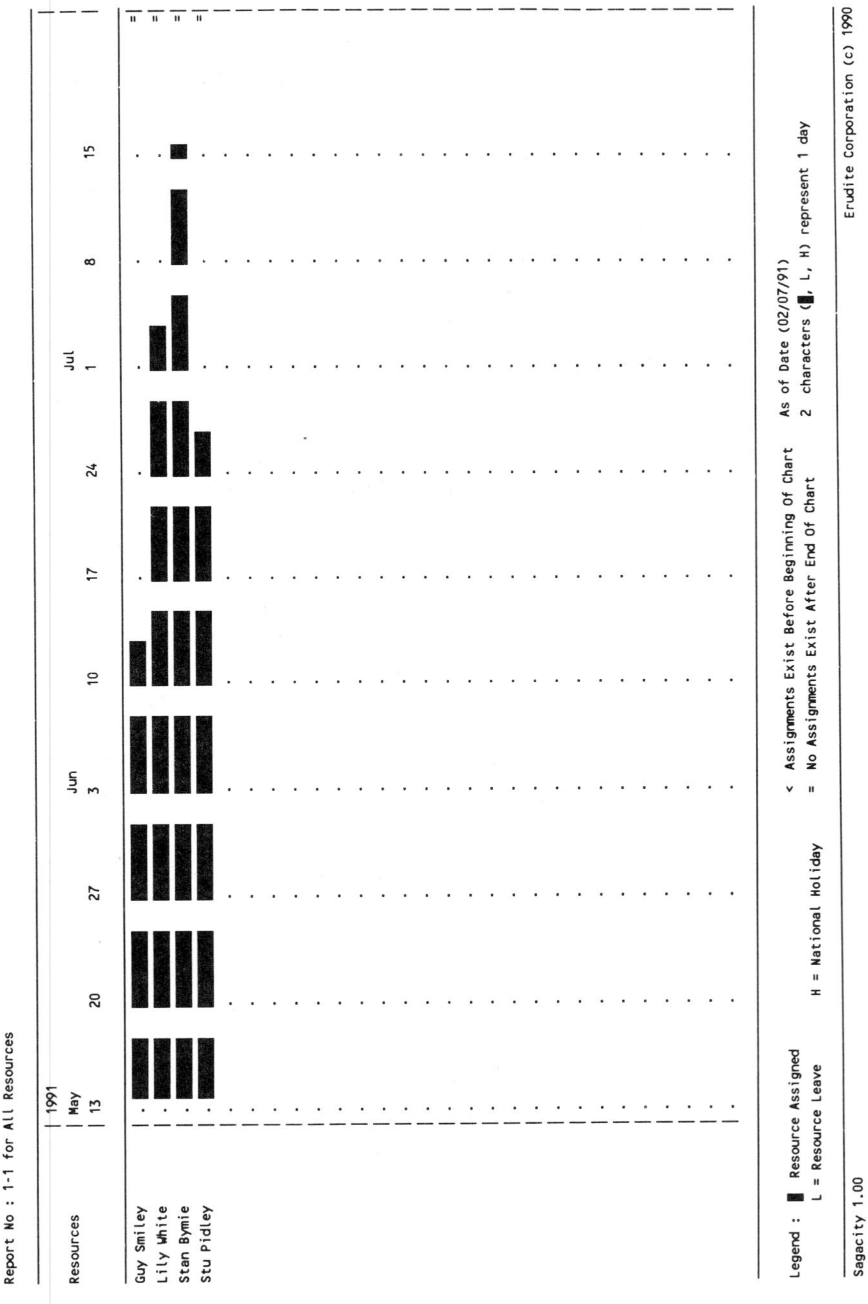

Report No : 1-1 for All Resources

Resources	1991 May 13	20	27	Jun 3	10	17	24	Jul 1	8	15

Guy Smiley
Lily White
Stan Bymie
Stu Pidley

Legend : ▉ Resource Assigned   H = National Holiday   < Assignments Exist Before Beginning Of Chart   As of Date (02/07/91)
       L = Resource Leave   = No Assignments Exist After End Of Chart   2 characters (▉, L, H) represent 1 day

Sagacity 1.00   Erudite Corporation (c) 1990

Figure 8.3   The Resource Summary Barchart shows the dates when resources are assigned or idle.

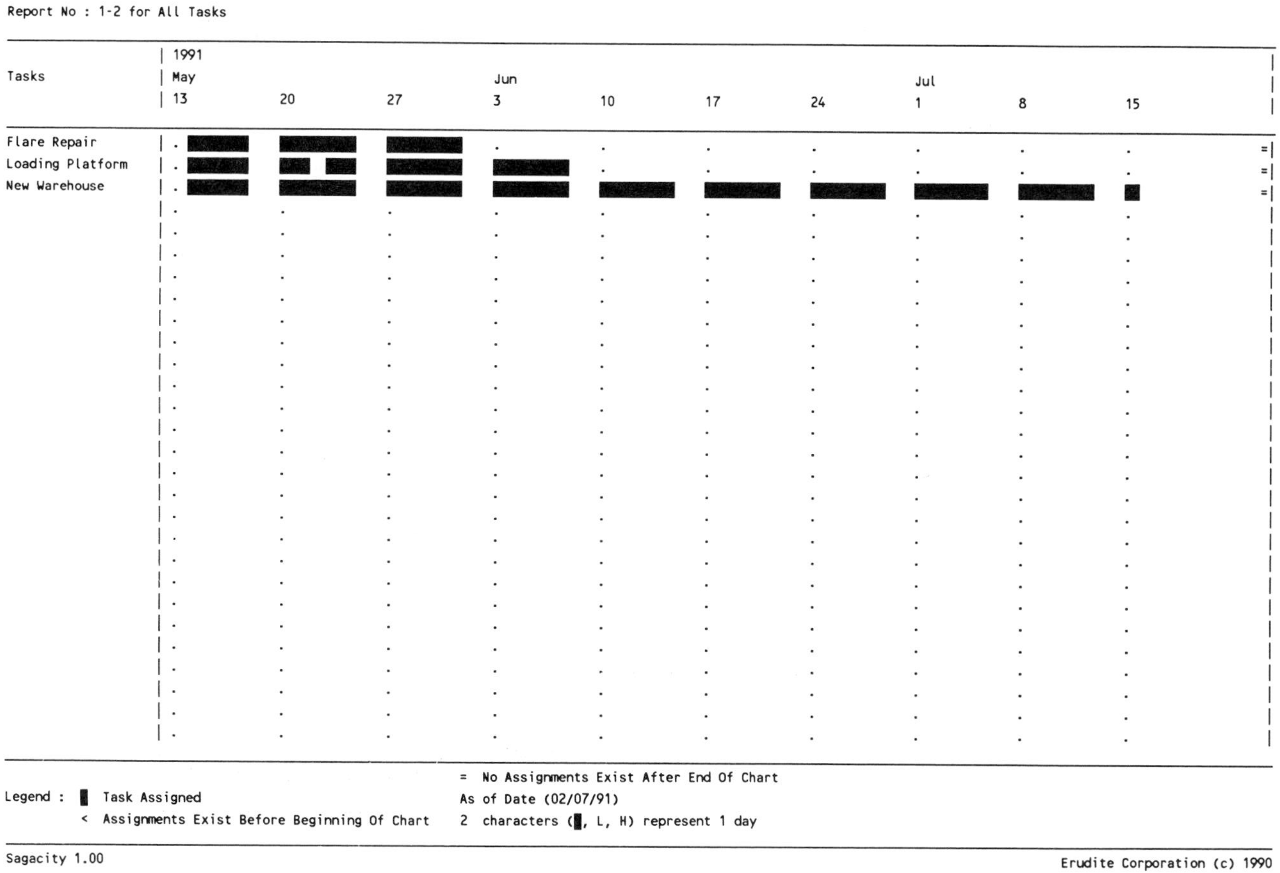

Report No : 1-2 for All Tasks

Figure 8.4 The Task Summary Barchart shows the dates when tasks are in progress, and indicates when an assignment will finish late.

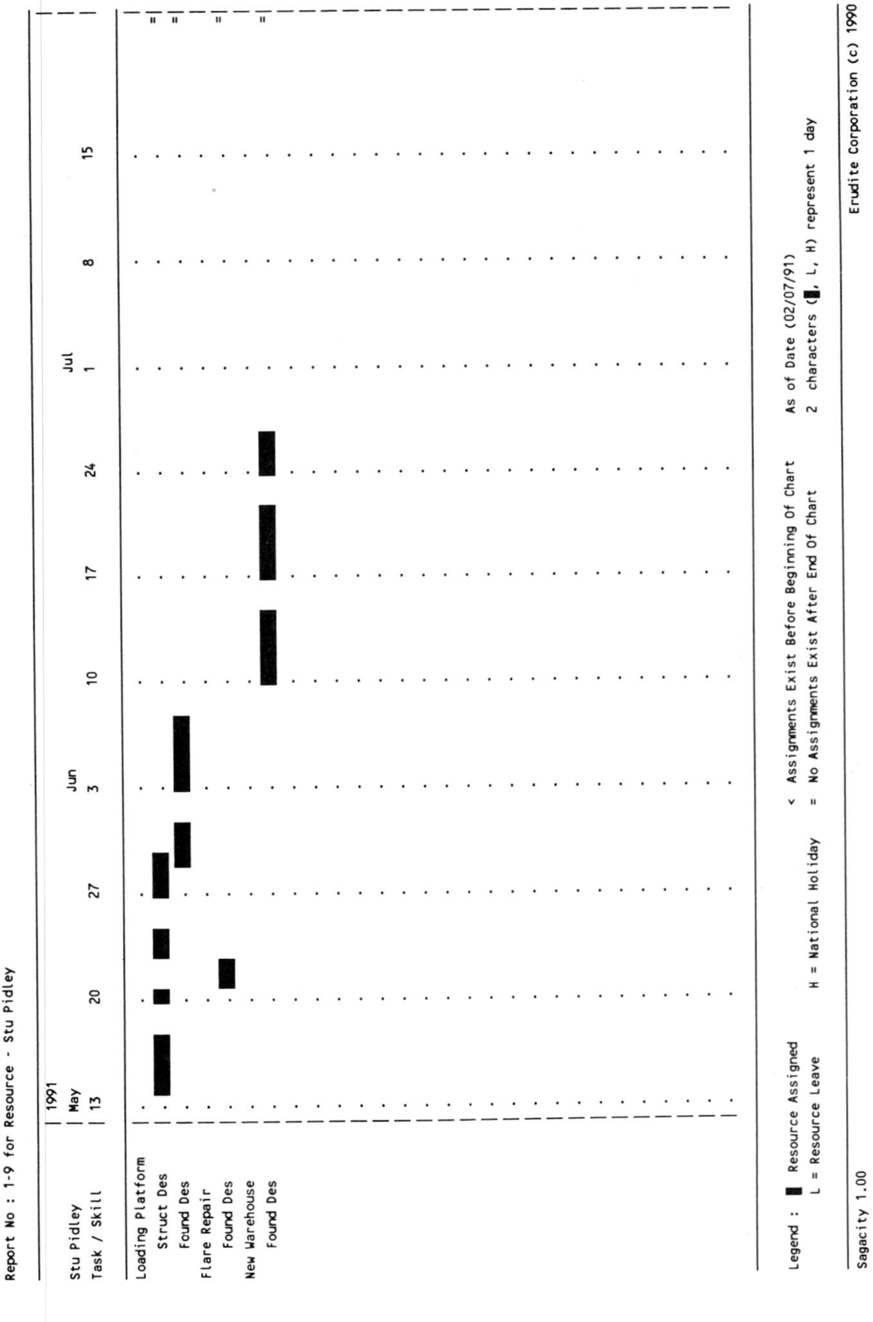

Report No : 1-9 for Resource - Stu Pidley

Stu Pidley	1991				Jun					Jul			
Task / Skill	May 13	20	27	3	10	17	24	1	8	15			

Loading Platform
Struct Des
Found Des
Flare Repair
Found Des
New Warehouse
Found Des

Legend : ▮ Resource Assigned   < Assignments Exist Before Beginning Of Chart   As of Date (02/07/91)

L = Resource Leave   H = National Holiday   = No Assignments Exist After End Of Chart   2 characters (▮, L, H) represent 1 day

Sagacity 1.00   Erudite Corporation (c) 1990

Figure 8.5   The Resource Detail Barchart shows what each resource is working on, and when.

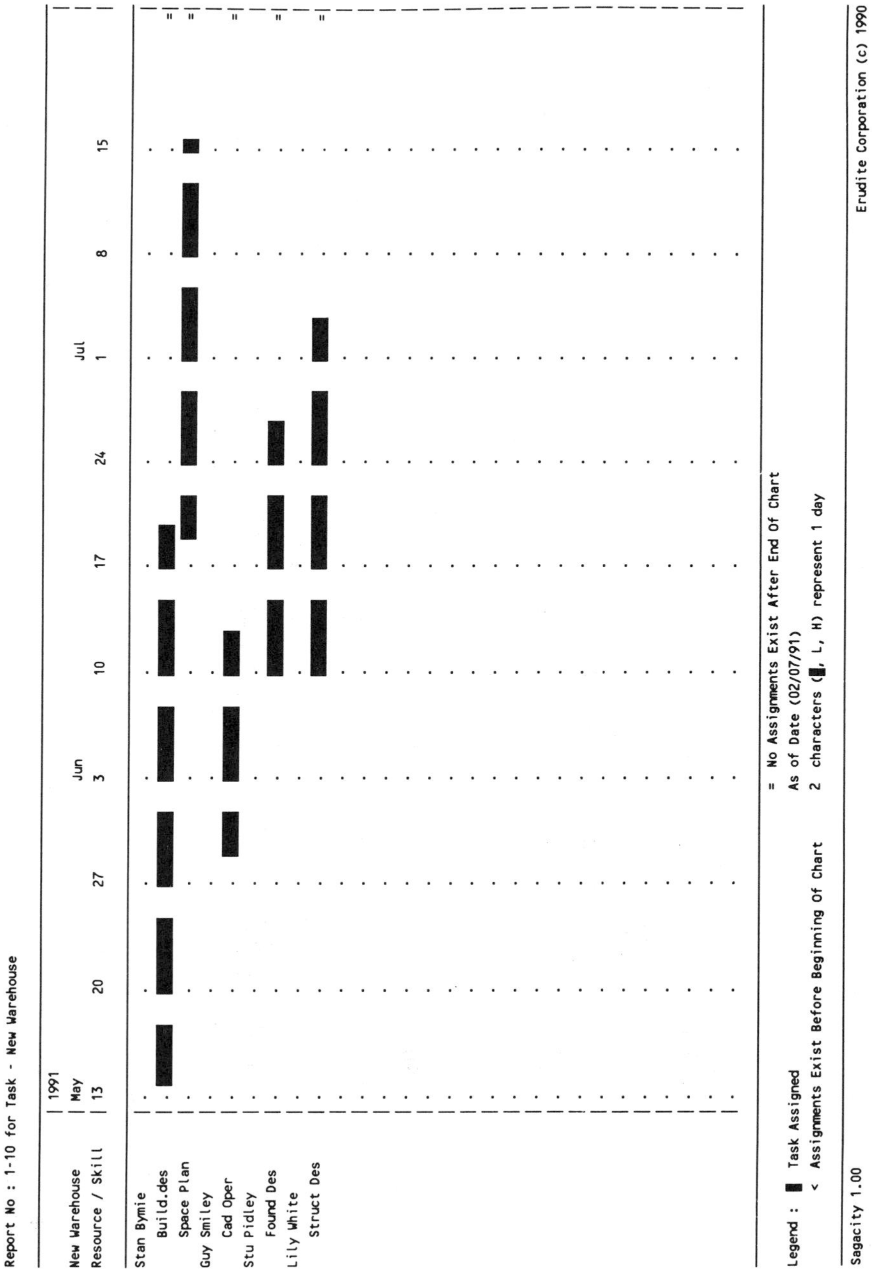

Report No : 1-10 for Task - New Warehouse

| New Warehouse | 1991 |
| Resource / Skill | May |

Stan Bymie
Build.des
Space Plan
Guy Smiley
Cad Oper
Stu Pidley
Found Des
Lily White
Struct Des

May 13 20 27 Jun 3 10 17 24 Jul 1 8 15

Legend : ▮ Task Assigned
< Assignments Exist Before Beginning Of Chart

= No Assignments Exist After End Of Chart
As of Date (02/07/91)
2 characters (▮, L, H) represent 1 day

Sagacity 1.00

Erudite Corporation (c) 1990

Figure 8.6   The Task Detail Barchart shows who is working on each task, and which skill they are using.

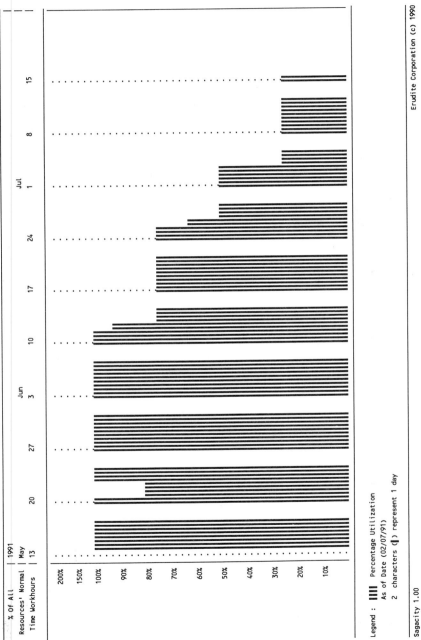

Figure 8.7   The Total Resource Utilization Histogram shows what percent of the available resource capacity is currently scheduled on tasks.

Report No : 1-3 for Resource Code - ENG1

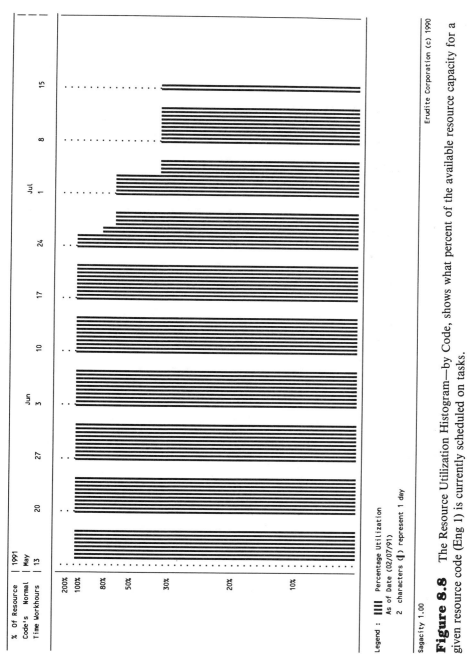

Legend : ▓▓▓▓ Percentage Utilization
          As of Date (02/07/91)
          2 characters (▐) represent 1 day

Sagacity 1.00

Erudite Corporation (c) 1990

Figure 8.8   The Resource Utilization Histogram—by Code, shows what percent of the available resource capacity for a given resource code (Eng 1) is currently scheduled on tasks.

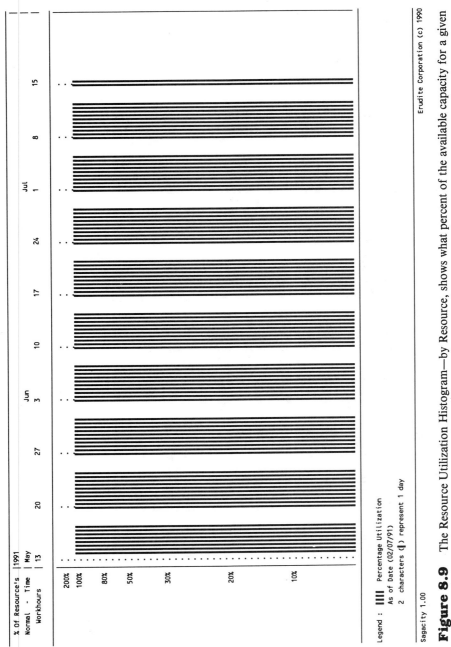

Report No : 1-3 for Resource - Stan Bymie

% Of Resource's |1991
Normal - Time | May
Workhours | 13

Legend : |||| Percentage Utilization
As of Date (02/07/91)
2 characters (|) represent 1 day

Sagacity 1.00

Erudite Corporation (c) 1990

Figure 8.9   The Resource Utilization Histogram—by Resource, shows what percent of the available capacity for a given resource (Stan Bymie) is currently scheduled on tasks.

Step 7: Interact with the Expert System for Further Optimization

Of course, no expert system can reflect all the considerations in the mind of a resource manager. For example, if it is not possible to complete all the tasks on time, overtime may be an option.

The Assignment Modeling method provides extensive opportunities for the resource manager to edit the input data and have the Expert System prepare a new assignment schedule. This process continues until we get an optimum solution that reflects all considerations.

Some variables that may be used for interaction are:

Allowing overtime to be specified with great precision by task or resource
Locking or unlocking resources from assignments
Granting or canceling leaves
Revising priorities or due dates

Step 8: Prepare Reports

The Assignment Modeling method provides tabular and graphic reports showing assignment scheduling information by task or resource.

CHAPTER SUMMARY

Effective resource scheduling is a key element in assuring successful projects and organizations. By using AI technology, the Assignment Modeling method creates easy-to-use software applications that solve the problem of finding an optimum way to assign tasks to resources.

For organizations using the Critical Path Method for task scheduling, Assignment Modeling provides a useful supplement for resource scheduling. Project managers can use PM software to define task schedules and effort requirements. This information can be communicated to resource managers who use Assignment Modeling software to assign resources to tasks and satisfy the requirements of all projects and project managers. At the time of this writing, the Assignment Modeling method is used in Sagacity[tm] software from Erudite Corporation. For organizations who have not found PM software to be useful, Assignment Modeling provides the focus on resource management they often need.

9

Organizing for Multiple Small Projects

THE EFFECT OF THE ORGANIZATION ON SMALL PROJECTS

In Chapter 1, project-management problems related to small projects are described. Most companies have some or all of the conditions described in that chapter including:

Priorities on the allocation of resources generally favor production
Lack of formal methods, procedures, systems, and reference data for use in planning and estimating
A lack of authority to match responsibilities of the project leader
Unclear lines of communication and responsibility
A lack of appreciation of the small project's difficulties due to their revamp aspects, the fact that many projects are managed at once, and the difficulties of working in an operating plant

It is evident from the above list that many of the problems of small projects are caused internally and therefore can be corrected internally. This chapter is written for project leaders or managers who have an interest in and an ability to change how things are done within the company. It discusses ways to strengthen the project-management function, as well as ways to remove obstacles to the effective performance of that function.

DEFINITION AND COMMUNICATION OF MANAGEMENT OBJECTIVES

General Objectives for Small Projects

One of the most frequent causes of conflict between project leaders and those who operate the plant facilities is the different interpretations of management objectives. Although maximizing profit is the clear overall objective, the way in which projects and operations each contribute to profit is often not clear. In the absence of guidance from plant management, the obvious contribution of production to profit, when compared with the overhead cost of engineering functions, tends to make any task that does not directly contribute to production appear a low priority.

This problem is, of course, nothing new. It is manifested most often in the allocations of plant resources. Since the timely application of the required resources is essential for projects to be completed on schedule and within budget, this problem can have a profound effect on the ability of the project leader to do his or her job.

The way in which small projects contribute to profit can be clarified by defining specific management objectives for each project or group of similar projects. Where necessary, guidelines should be developed for making decisions on resource allocations between projects as well as between project work and operations work.

Guidelines for Decision Making

On every project, large or small, the focus of the project leader is on schedule, capital cost, and design quality (not necessarily in that order). As we saw in Chapter 2, these parameters are directly related, and it is generally not possible to implement an improvement in one without affecting the others in a negative way. As a result, project-management decisions trade off one parameter against the other until the optimum combination is found. For example, a proposed design change adds quality by improving the maintainability of the unit. There is an investment cost to implement the change, as well as a schedule penalty. Should it be done? The answer lies in the relative importance to the company of maintainability, capital cost, and schedule. This relative importance depends on considerations of profitability which are by no means obvious. So, even in this simple and highly typical example, it is evident that, without some management guidelines, there is no way that a project leader or manager could make a decision and be sure that he or she had acted in the company's best interest.

Because few projects have well-defined guidelines for use in such decisions; because decisions like this are an every-day occurrence; and because the correct decision is usually impossible to ascertain by judgement, most project decisions are made without any way of knowing that they are correct. This, of course, can jeopardize the company's profitability. How, then, can project decision making be improved? One practical way is to issue guidelines for project decision making that specify the relative importance of:

Investment cost
Operating cost (including fuel consumption and maintenance)
Performance characteristics (e.g., service factor)
Startup date

These guidelines can apply to small projects generally or to a specific project. Examples of guidelines for decision making are:

It is worth $4,000 in increased investment for each day that the startup date is advanced
A 1% improvement in service factor is worth a 6% increase in investment cost
A 1% reduction in operating cost is worth 4% increase in investment cost

These guidelines can be easily derived from the economic analysis that was done to justify the project. Although the techniques of project-profitability analysis are outside the scope of this book, they generally involve calculating the projected cashflow for the project's operating life and discounting it back to present-day costs. It is therefore a straightforward exercise to determine the sensitivity of the projected profitability to variations in the variables of investment cost, time of startup, and operating cost. There are some cautions that should be heeded by users of these guidelines. Although they are useful, the guidelines can occasionally be abused by using them to justify irresponsible decisions. This is a problem that can be avoided by judicious monitoring and checking of the project decision-making process.

Life-Cycle Costs

The development of guidelines for decision making also introduces an important concept: that of "life-cycle costs." Life-cycle costs focus on the net operating cost of an item over its entire life, and include investment costs, energy consumption, maintenance cost, and operating efficiency. So, for example, if we specify a pump for a facility that is to operate for 30 years, we would normally select the pump with the lowest investment cost, as long as it met the design requirements of gallons per minute, pressure differential, etc. However, when life-cycle costs are considered, it might be apparent that a pump with a much higher investment cost would have to be shut down less frequently for servicing, be replaced less frequently, and perform more reliably resulting in a much greater profitability for the facility as a whole. The use of life-cycle cost analysis for design decision-making is discussed more fully in Chapter 10.

ESTABLISHING PRIORITIES FOR MULTIPLE PROJECTS

In the typical situation in which there are many small projects in progress at one time, there is often a lot that can be accomplished by simply establishing a system of priorities. In the absence of such a system, project leaders may be competing for resources and management attention in an unproductive way. For example, a critical project could be kept waiting while work is completed on a nonessential project. As in the case of guidelines for decision making, it is not always possible for the project leaders to assess the relative contribution of various projects to profitability. This requires a mangement perspective, and it is therefore management that must either set priorities on projects or establish a method whereby a priority can be established on each new project. Some of the questions one should ask in setting priorities include:

To what extent is the project essential? (i.e., one should distinguish essential projects from merely desirable projects, and establish levels of importance)

Does the project involve an essential system in the facility? (i.e., systems, such as electrical, firefighting, instrumentation, etc., should be ranked in terms of how essential they are to plant operation)

What is the financial attractiveness of the project? (i.e., one should distinguish projects with a satisfactory return from those with outstanding returns)

Using these considerations, we might establish a priority system as follows:

Workscope priority[a]	System priority[a]	Financial priority[a]
1	1	1
2	2	2

Workscope priority[a]	System priority[a]	Financial priority[a]
3	3	3
4	4	4
5 (lowest)	5	5

[a]1 = Highest.

The Net Project Priority is the weighted average of priorities in each category.

Along with the system of priorities, there should be procedures for implementing the system. This is desirable to avoid the situation in which only high-priority projects are given attention, and the lower-priority projects are ignored until their slippage becomes so great that their priorities have to be upgraded. This, of course, is an abuse of the system that should be eliminated so that projects of every priority can progress together, and one should need guidance only when conflicts in the allocation of resources arise.

PERSONNEL CONSIDERATIONS

Some of the problems of small projects stem from the perception that project work is less important than operations, and this should be addressed by changes in the way the project leader functions within the organization.

Continuity in the Project-Engineering Function

Some companies face the situation that the project function is used as a training ground for engineers who will, after a year or so, be transferred to other departments. Since most of the projects last less than a year, it sometimes seems that nothing is lost by this lack of continuity. A great deal is lost, in fact, and often the training that occurs is of questionable value.

Since small projects are in process continuously, some continuity in the project-engineering function is required to assure that something is learned from past projects and mistakes not repeated. Methods for planning, estimating and control, data, procedures, and systems should all be developed and maintained by the project-engineering function. If the project-engineering function had a core of experienced people whose main area of expertise was project work, and if proper methods, systems, and procedures were used in project work, then a year or two in project work would indeed be a valuable training experience.

Continuity in the project function also has the important benefit of fostering a sense of accountability. If projects are handled primarily by people who expect to be somewhere else in the near future, there is apt to be the feeling that mistakes will not become apparent until after the project leader has been transferred. The old refrain, "it can't be my fault, everything was fine when I left" only serves to diminish the sense of responsibility and can significantly weaken the entire project-management effort.

Authority and Responsibility

Another frequent concern of project leaders is that, though they are responsible for the successful outcome of the project, they have little authority with which to discharge that responsibility. Since they have to rely on other parts of the organization to provide resources (such as engineering services, drafting, purchasing, warehouse materials, and construction) they often must deal with those who are higher in the organization. This often makes it difficult to secure the necessary commitments to perform, and the necessary information for project control. While it would, of course, be inappropriate to make every project leader a manager, it is possible to define clearly the project leader's duties and responsibilities, along with the authority necessary to discharge those responsibilities effectively.

Project Leader's Performance Appraisal

Problems in Performance Appraisal

Appraisal of a project leader's performance is often difficult. Theoretically, one should judge performance in terms of whether most projects are completed on time and within budget. In the small-project environment this is not very easy, as small-project results can vary for many reasons such as scope changes, lack of access, interferences, and unforeseen difficulties, which are not the project leader's fault. And, if too much negative emphasis is put on schedule delays and cost overruns, the result is likely to be "fat" estimates and padded schedules, which, of course, do not enhance profitability in any way. What then can we do to establish performance criteria for the project leader? How can we measure his performance against these criteria?

Use of Standard Methods and Well-Defined Objectives

If the methods by which projects are to be managed are reasonably well defined, and management's objectives are clearly expressed in a system of priorities and guidelines, then the consistent behavior that sets the standard has been defined. The degree of skill, judgement, maturity, and creativity that project leaders display in the conduct of their work can then become the criteria for performance appraisal. The methods described throughout this book will form a satisfactory basis for such a process.

By defining the process, and measuring performance in terms of how well the process worked, we avoid the unfair judgements that often occur. Companies that punish project leaders who preside over an overrun, and managements that make it a career-threatening move to go back for additional funds without regard to the reasons why (and such situations are very common) only foster a climate of dishonesty. Many project leaders feel like the Greek messenger whose reward for telling bad news was to have his throat cut. That is an experience one does not wish to repeat, and it therefore encourages "padding" of future estimates and schedules.

If a company has a policy that states that estimates should have an equal chance of overrunning or underrunning (which it should), it is to be expected that some estimates will overrun by more than 10% and therefore need additional funds. It is a serious occurrence when a project overruns, but the questions should be:

Was the project leader effective in planning and controlling the job?
Was the original budget plan reasonable, based on the information available at that time?
Was the potential overrun revealed as soon as it became apparent, so that management action is still possible?
Are the reasons for the overrun things that could have been controlled, avoided, or compensated for by the project engineer?

Given a basis for a fair appraisal of this performance, most project leaders will not hesitate to report potential delays or overruns and, as we shall see in Section IV, the willingness to make realistic forecasts is one of the keys to effective project control.

Management Shares in the Responsibility

Successful projects are the result of effective cooperation between project personnel and company management. Management's role in a project is:

To establish the methods, systems, and procedures for project management

To establish and administer the organizational structure for projects

To approve the budget plan for each project

To review progress reports and take action when appropriate

The first two items describe management's role in creating the means to handle projects generally. These are described in "Establishing Communication" and "Developing Data and Methods for Small-Project Management," later in this chapter. The last two items describe management's role on each specific project, and are discussed below.

Approving the Budget Plan

The preceeding chapters have described the steps required to develop, present, and obtain approval of the budget plan. The budget plan, or project model, consists of:

The *network plan* (what is to be done)

The *schedule* (when it is to be done)

The *resource plan* (what it will take to do it)

The *cost estimate* (how much it will cost)

The *documentation* (of the design basis and important assumptions)

Management's approval of such a budget plan should mean more than "OK, go ahead." It should also imply agreement that the budget plan is consistent with the corporate objectives that the project is designed to meet, and that it has been prepared in accordance with established methods. Management, who have the ultimate responsibility, should be seen as senior partners in the project-management function.

Taking Action When Appropriate

There are apt to be conflicts, problems, and questions as the project progresses that cannot be resolved at the project leader's level. These should be referred to management and resolved promptly. Many projects are delayed due to the project leader's not being able to get an answer when one is needed.

ESTABLISHING COMMUNICATION

Multiple small projects, like most other corporate activities, benefit greatly from improved communication, particularly since so many diverse functions within the organization are involved. Some of the methods that have been successful in improving project-related communication include the following:

Project-Coordination Function

The project-coordination function consists of one or more people who serve as a coordination and communication channel between the project leader and the other functional groups. The coordination function helps relieve project leaders of that time-consuming part of their job. In the multiproject environment, the coordination function also helps to balance requirements between projects, thereby avoiding problems of resource allocation before they start. For example, if one project has slipped, the coordinator, who has an overview of all the projects, can easily see which project could best use the resources that have become available.

Small-Project Group

Some companies have found it worthwhile to set up a "small-project group." The purpose of this group is to integrate all the organization's functions related to small projects and, where possible, perform those functions itself. For example, the small-project group might have a small-project engineering function, so that a certain amount of straightforward design and drafting work could be done without recourse to the design and drafting department. Similarly, some purchasing and subcontracting functions could be placed within the group.

Status Review Meetings

Meetings to review the status of the multiple small projects, resolve problems, allocate work assignments and resources for the coming weeks, and advice management have proven to be worthwhile. If the project-reporting systems have been well defined, these can be used as the basis for discussion so that each project's status can be reviewed quickly.

Formal Reporting Procedures

In order to assure that management is kept informed, formal reporting procedures must be implemented. With the advent of personal computers this is much easier to accomplish than it was even a few years ago. Chapter 11 discusses management reporting in some detail, but the point to be made here is that the importance of sound and timely project reporting needs to be recognized by management, and the reporting effort supported by attention and prompt action when needed.

Encouraging an Enlightened Project Management Style

Most texts on management describe two general management styles:

Theory X

A management approach that assumes that those who are being managed dislike work, will seek to minimize the amount of work they do, have little ability to manage their own work, and little ambition.

In the case of Theory X, organizational goals are in conflict with individual goals, and sufficient effort to achieve organizational objectives can only be obtained by tight control and the threat of punishment.

Theory Y

A management approach that assumes that those who are being managed like to work, will work hard and exercise self-direction in order to achieve job satisfaction, will naturally seek responsibility, and will utilize ingenuity and creativity in solving problems.

In the case of Theory Y, organizational goals are more closely aligned with individual goals.

Most project management has a strong bias towards Theory X, especially where client vs. contractor relationships are concerned. In order to move to a working relationship based more on Theory Y, there have to be common goals as well as measurements of performance. Thus, the methods and systems for planning, estimating, and control which are discussed in this book can be thought of as a basis for better communication, and hence a means of gaining a project-management style more like that of Theory Y.

DEVELOPING DATA AND METHODS FOR MULTIPLE SMALL PROJECT MANAGEMENT

If effective project-management practices are to be implemented in an organization, the same integrated approach that works so well on projects should also be considered for methods development. If, as is so often the case, an organization tries to improve its performance in one area of project work, that improvement is apt to be diminished by a lack of improvement in other areas. So improvements are best addressed in a comprehensive manner that should include the following basic elements:

Databases for collecting, storing, and presenting data
Information coding systems

Planning and scheduling methods
Cost estimating methods
Guidelines for contracting
Quality assurance programs
Project control procedures

Each of these elements is described below.

Developing a Project Database

Why a Database for Multiple Small Projects?

A database can be thought of as a set of files of information organized in a useful way that provides the capability to present the data according to the user's requirements. A database provides specific, discrete information, as distinct from a method that provides generalized relationships and/or ways of operating.

Since almost all the data used in planning and estimating projects is empirical, the use of a database can contribute a great deal of consistency, accuracy, and credibility. Since most small projects occur in significant numbers, collection of data is relatively easy. So, setting up and maintaining a database for planning and estimating is a step that will greatly improve the small-project management function at minimal time and cost.

A database for project information also helps to assure that valuable project data is captured and made available for the next project. A classic problem with project data is that there never seems to be enough time to prepare final reports and see to it that the learning from the project at hand is documented and passed on. Too often, the pressures of the next project are too great, and the company is thus destined to repeat its mistakes. Fortunately, if the database is computerized, a lot of the data can be captured automatically.

Designing the Database

The most difficult aspect of a database is its design. To design a database, we must be able to specify the following:

The data required for planning, estimating, and resourcing. For example, what type of estimating do we wish to do? How much detail is required? Will we need piping costs for each flange and fitting, or will cost per ton of piping be sufficient?

The format that the data is generally available in. For example, do we generally receive field reports showing work-hours or do they show the number of people?

Other uses to which the data will be put. For example, can we use work-hour data that is already available in the accounting department?

The independent and dependent variables. For example, cost is usually dependent on independent-design variables such as pump horsepower, tons of steel, etc. How should they be related in the database?

The number and relationship of variables that are to be handled at once.

The facilities that will handle the database. Will we use mainframe or personal computers? What limitations exist as to volume of data handled?

The design of a computerized database must also specify how data is to be processed, in terms of input, internal processing, and output (visual displays, printed reports, tape or disc storage). There are three distinct types of structure for a computer database:

Hierarchal database: This locates information in progressive levels of detail. For example, a post office zip code locates the area of an address by progressively defining state, city, area, and group of streets.

Relational database: This locates information by relating "key fields" in one database to another.

Network database: This allows a "many-to-many" relationship of data fields.

As we extract information from various files in the hierarchal database, the system proceeds down through the hierarchy each time. A relational or network database allows us to extract data from each file without using the hierarchy. When evaluating computerized databases, the ability to select, sort, aggregate, arrange, and display data the way we want to is the most important feature, and the database structure helps determine how much flexibility we will have.

Setting up the Database

There are many good programs for data management, including those for personal computers. For small projects, these simple and inexpensive databases work fine. Often, the existing in-house computer can be used. This is particularly appropriate when input data is already available in the system.

When setting up the database it is essential to have an information coding system to organize and recall the data. This is discussed in the next section. The database must be on a consistent basis throughout, and that basis must be clearly defined in terms of the conditions assumed, performance by client and contractor, pricing and productivity levels, "base date" for costs, etc.

A database design often consists of several databases that can be used separately or together. For example, if the integrated approach is to be used for planning, resourcing, and estimating, the databases of work-hours, durations, and costs should all be related to assure that consistent results are produced.

An important function of a database is to collect raw data from projects and to use this data to assure that the database is up-to-date. If, however, raw data was fed directly into the database, there would be a lot of confusion as a result. Raw data must first be analyzed to put it on the same time-and-cost basis as the rest of the data in the database. This process of normalization requires judgement, and a knowledge of the circumstances surrounding each project. A simple data system design is shown in Figure 9.1, illustrating the *holding database* in which raw data is stored, as well as the *reference database*, which is used for planning and estimating. The reference database should be protected from unauthorized modification, and should be updated every 3 to 6 months, based on analysis of data from the holding database.

Sources of Data

The quality of the data is, of course, the main consideration of a database and, in the initial stages, can be a problem. Bearing in mind that the best data is that which best predicts what will happen on a new project, it can be seen that historical data is not always the most desirable. For example, we may be aware that the last 4 turnarounds were on schedule, but used excessive work-hours and overran their budgets. Should we use data from these projects in planning and estimating the next one? To do so would imply that the unsatisfactory past performance is something we expect to repeat. Not to do so could be interpreted as a form of "cockeyed optimism," which has no place in budgets.

Since our purpose is to predict that outcome which has a 50/50 chance of being achieved, we should predict, with our plan and estimate, that the current performance will be better than the past if that is what we expect to happen. To make that judgement, we should know why the past projects encountered problems, and why the current one can be expected to avoid these problems. Having determined those answers, we can make adjustments to the data and thereby establish the basis for future estimates. However, if analysis of the past data shows that we cannot reasonably expect the performance to be improved, then the unaltered past data is the best predictor of future performance.

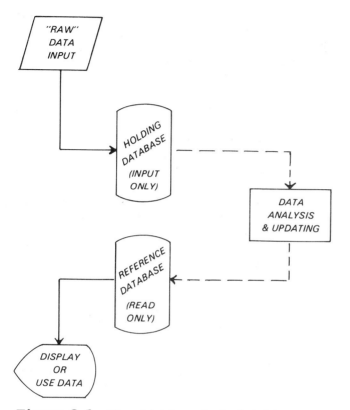

Figure 9.1 Use of databases for project data.

The above illustrates the amount of work that goes into defining and loading data. An additional source of work is obtaining the data in the first place. Unless past projects have been managed with an eye towards collection of data (and how many are?) we are likely to find that the collection, organization, analysis, and normalization of data can be a daunting proposition. It is important that it be done right, since all future projects will be based on the data we put in. Although the preferred method is to use actual past data, even if it is adjusted, there are other ways to load the database if the use of actual data turns out to be impractical. Many companies have solved this problem by accessing a database that is available commercially. Some of these databases are quite good, and there are a number available for architectural applications as well as process-industry projects. The problem is that the industry average may not fit the typical small project in a given company. Although this can be easily addressed through the use of correction factors, it takes some time to establish those factors by comparing actual vs. predicted results.

One approach that works well is the use of "synthetic data." This involves creating data that represents what is expected, even if it does not correspond to actual past data. At worst, synthetic data will give results that are about what would have been estimated anyway. As actual data is collected, normalized, and used to update the database, the synthetic data is replaced by actual data. To develop synthetic data, standard or typical designs are produced for a variety of design conditions. The expected cost is estimated for each case. Similarly, standard networks are prepared and resourced for each design case. The data is then broken down into the detail required to fit the database format.

Designing the Information Coding System

What Is It?

The information coding system organizes the flow of project data. Although it might, at first, appear to be a simple task, the design of a coding system for project information is quite difficult, particularly for small projects in an operating plant, because so many different organization functions are involved. The functions that deal with project data include:

Contract administration
Purchasing
Planning and scheduling
Cost engineering
Project engineering
Accounting
Field supervision
Documentation
Quality assurance
Company management

On small projects, some persons may perform more than one function. The coding system gives each person and each function the information he needs, when he needs it, in the format with which he is comfortable.

Coding System Requirements

The project-information coding systems should be:

Simple to use (i.e., long, complex codes are to be avoided)

Compatible with the company's codes and those used by contractors

Flexible enough to sort, select, and present information according to user requirements

Compatible with the constraints imposed by the computer system to be used

Capable of handling an integrated approach to planning, estimating, and resourcing

To meet these requirements, coding systems often function in a hierarchy of codes and subcodes, such that excess detail is not a problem to each user. A work breakdown system (WBS) is one way to accomplish this.

Elements of the Information Coding System

The basic elements on which the coding system is based are:

The network plan

The cost coding system

The organization chart

The work breakdown structure (if applicable)

The design coding system

The accounting system

The coding system manages information according to the specifications of these basic elements. It allows us to manipulate the project model, and prepare reports using cost data, resource data, or scheduling data. It should work as follows:

The network plan should have a code for identifying activities for use in scheduling, resourcing, and progress measurement

The resources should have a cost code, identifying them by labor craft, material type, etc.

The organization chart specifies who gets what information and in what format

The work-breakdown structure (WBS) relates the network activities to specific contracts or work packages. This technique is useful in the multiple small-project environment for selecting and collecting data.

The design code should be used to indicate what system an item of material belongs to, or what type of material it is.

The accounting code must be compatible with the cost data received from the project.

By examining each of these elements, and combining them wherever possible, a practical coding system can be developed. A simple but effective coding system concept is illustrated in Figure 9.2. The coding system as seen here is a matrix in which the rows are physical components of the project, and the columns are the activities (or tasks) that are performed on those components. The cell, then, describes a "cost center" that relates the physical component to the tasks of the project. Should additional detail be required, the matrix can be extended into additional dimensions.

Developing a Method for Planning and Scheduling

The technology of planning and scheduling was discussed in Chapter 3. To implement such techniques in-house, an organization has to first define its needs and constraints so it can develop specifications for its method. Some of the points which should be considered are:

The specific organizational needs that the method must satisfy
The specific methods to be used

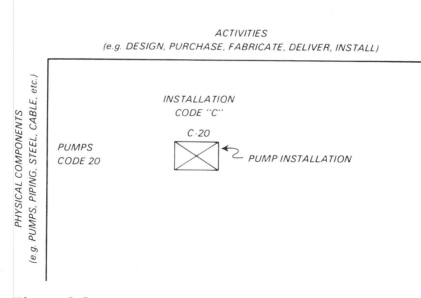

Figure 9.2   Basis elements of a project-information coding system.

The formats and nomenclature to be used (i.e., the definitions of levels
of networks and when each should be used)
The data required (we must assure that the database will provide it)
The use of standard networks for similar projects
The procedures for preparation and review
The amount of training required
The computer systems that are available or will be required
The interfaces between the planning and scheduling methods and those
for resource planning, cost estimating, and project control.

Having defined the specification for the method, it can be developed
and implemented in the following steps:

1. Establish database and coding system (see sections above)
2. Prepare standard or sample networks for training
3. Define computer requirements, and select and implement system (see
   Chapters 14 and 15)
4. Prepare procedures
5. Select the first project to be planned with the new method
6. Plan the first project using the new method in parallel with existing
   methods
7. Evaluate the results, and modify as required
8. Obtain management approval for implementationn
9. Train users

Developing a Method for Cost Estimating

As in planning methods, the development of a method for cost estimating begins with a definition of company needs so that one can define a
specification for the method. The steps that should be taken are shown
below.

Definition of Estimate Stages

Typical stages of a project during which estimates are prepared are:

Screening, during which projects are compared and some selected to progress further
Design optimization, during which the cost of alternative designs are
studied
Project approval, during which the approved budget is established

It is evident from the above that, in the case of the screening and
optimization estimates, relative costs are more important than absolute

costs. The estimate prepared for project approval, however, must be as accurate as possible. Small projects in some companies are in simpler situation of preparing only one estimate: the budget estimate.

Definition of the Estimating Basis at Each Stage

In order to know what estimating method will be most appropriate at each stage, the degree of definition of the estimate basis should be defined at each stage. Consideration should be given to:

Design: To what extent has the design progressed at each stage? How much of the design basis is apt to be preliminary? How much will be firm?

Planning: How well defined are the plan and schedule likely to be?

Cost data: What type of data is likely to be available?

The estimating method must be appropriate to the basis definition as well as to the use to which the estimate will be put.

Definition of Estimate Use at Each Stage

Estimates help in decision making in a number of different ways, including:

Deciding which projects are worthy of further work, developing preliminary budgets, and performing preliminary economic analysis (at the screening stage)

Deciding which design or planning approaches are most desirable, refining budgets, and obtaining approval to proceed with further design work (at the optimization stage)

Obtaining project approval, budgeting, and setting up cost control procedures (at the approval stage)

Definition of the Required Accuracy at Each Stage

Based on the use to which the estimate will be put, we can define the accuracy required for each type of estimate. The accuracy that can be achieved is, of course, only as great as that of the basis. It is important to recognize that estimate accuracy cannot exceed the accuracy of the basis, and therefore reasonable expectations of estimate accuracy are in order.

Development of Design-Cost Correlations

The cost-estimating method is intended to provide generalized relationships between cost (the dependent variable) and independent variables such as design quantities. Once the database is established, the method, either in the form of estimating curves and tables or computerized equations, can be established through the development of cost correlations. The design of the method, i.e., what correlations are established and their statistical accuracy, should match the specifications developed in the above steps.

An important point in methods development is that a certain amount of inaccuracy is acceptable. Many estimators only feel comfortable with a lot of detail and it is a common mistake to prepare estimates in excrutiating detail that contributes little to overall accuracy. This can be easily seen in the case of estimates that are escalated over significant time periods: the escalation factors are bound to be relatively inaccurate, yet the escalation cost may be as large as the base estimate itself. So the extra effort required to add accuracy through increased detail (which increases estimating time and cost) should be evaluated carefully to see whether the added accuracy is significant and necessary.

Documentation and Training

As in the planning method, final steps in implementing a planning method include documentation, preparation of procedures, management approval, training, and testing on a sample project.

Guidelines for Contracting

The various approaches to contracting are described in Chapter 3. It becomes evident there that there is a good deal of judgement involved in selecting the right approach for a particular project and for the company in general. Because contracting decisions can involve considerations, such as legal aspects, that are outside the realm of most project engineers, it is important that management guidelines be provided. These guidelines should indicate what type of contract is preferred for each type of situation, and what factors should be used to decide on a contracting approach.

Once the project leader has selected the type of contract to be used, the guidelines can also provide typical contract forms that have been reviewed by the legal staff and approved for project use. This avoids any dangerous legal implications from the blunders of inexperienced project engineers, and streamlines the approval process.

Procedures for Quality Assurance

A complete set of project procedures must include quality assurance. As discussed in Chapter 3, the quality assurance program must be well planned. Like contracting, decisions regarding design quality (see ''Guidelines for

Decision Making,'' earlier in this chapter) often require an appreciation of management considerations that are not generally available to the project engineer.

The quality-assurance procedure should be the result of an in-depth design and cost analysis, based on life-cycle costing as discussed in ''Life-Cycle Costs,'' earlier in this chapter. The first priority is to define quality in specific design terms, such as those used in reliability engineering (e.g., ''mean time between overhauls,'' and ''mean time to replace''), as well as those relating to operating efficiency and maintainability. Once these specific design requirements are defined, they can be included in the bid package for equipment thereby assuring that the desired quality is obtained.

Quality assurance also requires an effective change-order procedure, since many design improvements are brought into the job as changes. If changes are properly evaluated, according to guidelines as discussed in ''Guidelines for Decision Making,'' the optimum blend of design quality and investment cost can be achieved.

It should be remembered, when discussing quality and cost, that the design of a project is the aspect over which we have the most control and that design has the most profound effect on cost. Therefore, any procedure that helps assure that the cost implications of design decisions are considered will go a long way to assuring a profitable project.

Project-Control Procedures

The principles of project control are discussed in detail in Part IV. It is, however, appropriate to include project control in this discussion, as it is the means by which management objectives are achieved. In fact, everything that has been discussed in the first 9 chapters of this book provides a proper basis for project control, i.e., organizations, methods, data, systems, and procedures for planning, resourcing, estimating, quality assurance, and contracting. We now have something against which we can measure: a comprehensive, integrated plan.

If company management has provided the tools for planning and budgeting the project, it will also want to make sure that the systems and procedures are in place for controlling it so that the desired results are achieved. The steps that are required to develop and implement project-control procedures include:

The definition of the objectives (i.e., what each procedure is to accomplish)

The description of the method (i.e., what basic techniques are to be used)
The allocation of responsibility (i.e., who does what to whom)
The definition of the flow of data and information
The preparation of an estimate of time and cost to implement
The gaining of management approval
The development and implementation of computer systems (if needed)
Documentation
Testing on a sample project in parallel with existing procedures
Modification as required
Training
The briefing of contractors and revisions to contract documents
Complete implementation

CHAPTER SUMMARY

This chapter covered management's responsibility in the successful completion of small projects. Much can be done internally to assure that the necessary data, methods, systems, and procedures are available to the project leader. Some simple modifications to the organization can help a great deal as well.

With company management and project engineering now armed with a comprehensive, integrated plan, a schedule and estimate, and an approved budget, we are ready to discuss the part most of us enjoy the most: the execution of the work.

IV
EXECUTING THE SMALL PROJECT

10

Optimizing the Design

DESIGN OPTIMIZATION: A POWERFUL TOOL FOR PROFITABLE PROJECTS

What is the purpose of a company project? Of course, the answer is profitability. Paradoxically, few projects are managed as if profitability were the objective. Instead, most companies calculate profitability in order to get the project approved, then simply forget about it.

The unfortunate result of current project-management practice on many small projects is that most project decisions are subsequently made without regard to profitability. Instead, they are made with a short-range perspective that may result in the wrong design decision.

Let us explore the ways in which design decisions affect profitability and how the design decision process may be improved on small projects.

DESIGN DETERMINES A PROJECT'S OUTCOME

What determines the outcome of a project? Common sense tells us that it is:

The design of the facility or product
How well the project is managed
The effect of external factors on the project

It is interesting to note that, of these factors, the most powerful one is the first: the design.

The design of a product, system, or facility determines the minimum cost to build or develop it, and also determines the profit it will generate for the company. Once the design is set, other factors come into play, but these are less significant.

The design phase of any project represents the best opportunity we will have to do work and make decisions that have a profound impact on the project's success. So why do so many companies rush through the design process? In so doing, they are throwing away this important opportunity to truly optimize the project's results.

Design optimization is the process by which the design work can be planned and conducted in such a way that an optimum design, one that maximizes the profit returned by the project, is reached.

PROFITABILITY: THE OBJECTIVE OF EVERY PROJECT

Just about every project shares the same objective: to make a profit for the company. Organizations invest in projects for the same reason they invest in anything else, and the small-project manager's job is primarily to assure that the project is designed and executed in such as way as to maximize the return on that investment. Yet few projects are managed as if profitability were the objective.

Profitability Is Determined by Project Decisions

The performance of most project managers (large or small) is measured by whether the project finishes on or under budget. Although the investment cost is an important variable in determining profitability, it is by no means the only one. Figure 10.1 shows the variables that determine a project's profitability. These are:

Investment cost
Date at which the project starts producing revenue
Capacity of product produced
Specification and price of the product

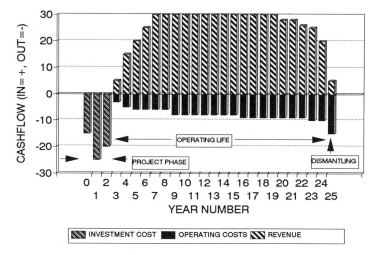

Figure 10.1 Design decisions affect a project over its entire operating life. Life Cycle Costs are the Net Present Value of all the cashflows from the start of the project to the dismantling of the facility.

Cost to operate the system or facility—consisting of operating, maintenance, utility costs.

It is interesting to note that, of these variables, *all are affected by decisions made during the relatively short project phase.*

So one might ask whether the practice of making design decisions and evaluating project performance based on minimizing the investment cost is really correct. If the goal of a project is to create profit, then a decision criterion that ignores the profitability of alternative decisions cannot be right. In fact, it is easy to see how the design alternative that gives the lowest investment cost may well be one that is less energy efficient, less reliable, more difficult to maintain, and therefore more expensive in the long run.

Design Decisions Should Be Based on Life-Cycle Costs

Figure 10.1 also illustrates that a "project" actually has a very long life. What we usually think of as the "project phase" (covering design and execution) is really only a brief period in the total life of the facility. After it is built, the system or facility will (hopefully) produce revenue and profit for the company for many years to come. At the end of its useful life, it will be dismantled at some cost that may be offset by some salvage value.

When one thinks of project decisions in this way, it is clear that many project managers make decisions that are short-sighted. If design decisions are made with the objective of minimizing the investment cost, higher operating, maintenance and utility costs may well result. It is clear, therefore, that investment cost is not the best way to evaluate the "cost" of a design approach. *A much better way is with life-cycle costs.*

The life-cycle cost of a facility, system, or component, is the present-day cost to design, build, and operate and maintain it over its entire useful life, plus the cost to dismantle it at the end of its life (net of any salvage value). In other words, life cycle cost is the net present value of:

Investment cost
Annual operating costs
Maintenance costs
Cost of utilities
Cost to dismantle—offset by salvage value

For example, a net present value life-cycle cost of $1 million means that the company would have to put $1 million in the bank, at current rates of interest, in order to have enough money to pay for designing, building, and operating the system, facility or component for its entire operating life, as well as the cost to dismantle it at the end. If we refer to Figure 10.1, we can see that the Life Cycle Cost (LCC) is simply the Net Present Value (NPV) of all the costs on the lower half of the curve.

Calculating Life Cycle Costs

The calculation of Life Cycle Costs is a simple process. To begin with, one must understand the concept of Net Present Value. The Present Value (PV) of a series of expenditures is simply the equivalent amount of money that would be set aside today in order to have enough funds to cover all the expenditures when they occur. To calculate PV, we must also know the prevailing interest rate representing the cost or value of the company's money with time. The calculation is then performed using tables, a calculator with "business" functions, or a spreadsheet program such as Lotus 123®, Quattro® or Excel®.

A typical PV calculation might indicate that the present value of an expenditure of $10,000/year, starting one year from now and continuing for five years, would be \$37,908 if the time-value of money was 10%. Note that the algebraic sum of the payments, \$50,000, is much more than the discounted present value.

The Net Present Value calculation is similar but it allows us to calculate each year's expenditures as a net value of positive and negative amounts. For example, we might estimate each year's expenditures for a process unit as follows:

Operating costs (staff, supplies etc.)
Plus utilities costs (e.g., power consumption)
Plus maintenance costs
Plus other costs (e.g., insurance, overheads, user fees, licenses, etc.)

The NPV calculation adds these costs together for each year, then calculates the present value of that year's costs.

Spreadsheet software, such as Lotus, Excel, or Quattro, is an excellent way to perform LCC calculations. Macro-driven applications can be set up using the program's "NPV" functions so that the input of data and preparation of reports is simple and fast. This technique is illustrated by the following figures:

Figure 10.2a shows an input screen for descriptive data
Figure 10.2b shows an input screen for investment costs
Figure 10.2c shows an input screen for operations and maintenance costs
Figure 10.2d shows an input screen for salvage and dismantling costs

PROJECT NAME: [   ]

ENGINEER: [   ]

INTEREST RATE: [   ] (enter as a decimal, eg .10 for 10%)

DESCRIPTION: [   ]
                        [   ]
                        [   ]

Figure 10.2a   General Information Input Screen for Value Engineering Analysis.

MATERIALS		LABOR			TOTAL
DESCRIPTION	COST	MANHOURS	COST/MH	COST	COST
PIPING	$375,000	1500.0	$75	$112,500	$487,500
ELECTRICAL	$155,000	2500.0	$50	$125,000	$280,000
PUMPS	$75,000	800.0	$125	$100,000	$175,000
				$0	$0
				$0	$0
				$0	$0
				$0	$0
				$0	$0
				$0	$0
				$0	$0
				$0	$0
				$0	$0
				$0	$0

Figure 10.2b   Input Screen for Investment Costs.

	YEAR				
CATEGORY	1	2	3	4	5
MAINTENANCE COSTS	5500	5500	7000	6000	5000
OPERATING COSTS	1500	1600	1600	1700	1800
ENERGY COSTS	600	600	700	700	750
OTHER COSTS	500	600	600	600	600
TOTAL	8100	8300	9900	9000	8150

Figure 10.2c   Input Screen for Operating and Maintenance Costs.

SALVAGE VALUE IN LAST YEAR: $5,000

COST TO DISMANTLE & REMOVE: $25,000

NET COST/INCOME @ END $20,000

Figure 10.2d   Input Screen for Salvage Value and Dismantling Costs.

DESIGN CASE:	NET PRESENT VALUE LIFE-CYCLE COST	RELATIVE LCC	DESCRIPTION
BASE	$1,005,950	100%	STANDARD DESIGN
ALTERNATIVE I	$1,200,675	119%	NEW PUMP CONFIG.
ALTERNATIVE II	$950,843	95%	REDUCED PLOT AREA
ALTERNATIVE III	$754,332	75%	LOAD SHARING

Figure 10.3a   Summary of Life Cycle costs for Base Case and each alternative.

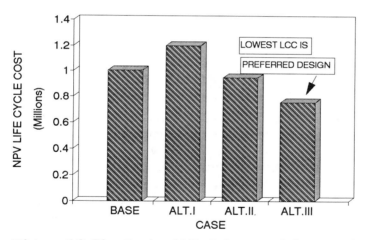

Figure 10.3b   Results of Life Cycle cost analysis presented graphically.

LIFE-CYCLE COST - NET PRESENT VALUE CALCULATION

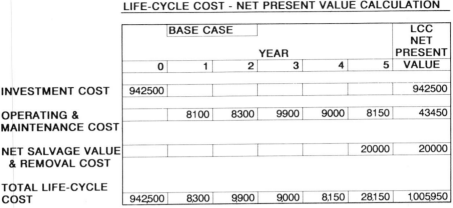

	BASE CASE						LCC NET PRESENT VALUE
				YEAR			
	0	1	2	3	4	5	
INVESTMENT COST	942500						942500
OPERATING & MAINTENANCE COST		8100	8300	9900	9000	8150	43450
NET SALVAGE VALUE & REMOVAL COST						20000	20000
TOTAL LIFE-CYCLE COST	942500	8300	9900	9000	8150	28150	1,005950

Figure 10.3c   Life Cycle cost analysis for the Base Case.

Figure 10.3a shows an output screen summarizing the results
Figure 10.3b shows a graph of the results
Figure 10.3c shows an output screen showing the LCC calculation for
    one alternative

ELEMENTS OF DESIGN OPTIMIZATION

Design optimization consists of the following elements:

Assuring the technical validity of the design work—i.e., the compliance
    with standards and good engineering practice
Controlling the time and cost of the design work
Creating an optimum design that maximizes project profitability

The first of these needs is met by the processes of design checking that are
well-known and outside the scope of this book. The second need—to control
the time and cost of the work—is covered by Chapters 11-13. That leaves this
chapter to discuss the third, and most important need for design management: to create an optimum design that maximizes project profitability.

Design Optimization—Part of the Quality Management Process

Quality Management is a concept of high priority for many, if not most,
successful corporations, and rightly so. In fact, design optimization is an

important part of quality management as it applies to engineering and construction projects.

Most accepted definitions of quality stress the concept of "conformance to requirements." That is, a quality design and product is one that meets or exceeds the customer's requirements. This is an interesting concept for designers, as it raises several questions:

1. Who is the real "customer" whose requirements must be met?
2. What is the right set of requirements that the design should meet?
3. Is the customer always the best one to define requirements?
4. How can essential requirements (i.e., needs) be distinguished from nonessential requirements (i.e., wants)?

Design Optimization Means Questioning Requirements

Let us review each of the questions posed above:

1. *Who is the real "customer" whose requirements must be met?*

Consider an engineer at the XYZ Chemical Manufacturing Company who has been given a project to design and construct a process unit to make a new product. Who is the "customer" in this case?

The ultimate "customer" is, of course, the consumer who buys a product made from the chemical raw material that the facility built by this new project will produce. This "customer" is represented by the Marketing Department who has listened to the consumer and identified the specifications for a product to meet the consumer's need. So, within the XYZ Company, Marketing is the customer.

Or is it? The process unit will be operated by Plant Operations. They are also a customer. It will be maintained by the Maintenance Department—they are also a customer. It will be constructed by the Construction Department—they are also a customer of our project engineer. Finally, the company's stockholders are a customer as well, as they are represented by management and also have requirements.

All of these customers have requirements, which consist of essential needs and nonessential wants. And their requirements may not always coincide. For example, the stockholders may feel that building the project for a minimum investment is a primary requirement. This may mean that money may not be made available to meet requirements of other customers, such as those who will operate and maintain the plant. If all customers were allowed to get everything they wanted, any project would be impractical and unprofitable. So, defining requirements properly is an important first step to being able to satisfy them.

2. *What is the right set of requirements that the design should meet?*

As discussed in the preceding paragraphs, it is not always apparent what the right set of requirements should be. One difficulty in defining requirements is separating wants from needs. Therefore, an important step in design optimization is to begin the design process by carefully evaluating the requirements that the design should meet. It may also be appropriate to evaluate requirements that are clearly "wants" (i.e., nonessential) in terms of what it will cost to provide them. The decision may well be that the cost is worthwhile.

3. *Is the customer always the best one to define requirements?*

Design Optimization begins with the questioning of requirements. Why question the customer's requirements? Isn't the customer the one who knows the requirements best?

Actually, the answer is no. In the corporate environment, there are many customers with different and conflicting requirements. Furthermore, customers are often the least qualified to know what the right set of requirements are.

Customers may overstate their requirements for several reasons:

They may be unaware of the technical implications of some requirements, or the cost of implementing them

They may be unaware of alternative methods with which their requirements might be met

They may simply not have enough time to work on specifying requirements, so they play it safe and overspecify

For example, a customer for a laboratory building might be a research scientist. The scientist may know what lab requirements are essential, but not know their relationship to requirements for electrical, architectural and mechanical facilities. She may not realize that certain requirements are costly, whereas others are not.

4. *How can essential requirements (i.e., needs) be distinguished from nonessential requirements (i.e., wants)?*

Ideally, the designer and customer should be partners in working to define the right set of requirements. The designer can advise the customer of the life-cycle costs of various requirements. The designer can then propose alternative methods that maintain quality (i.e., meet essential requirements), while minimizing life-cycle cost.

An important function of design optimization is, therefore, to begin the design process by questioning requirements. To the design function falls the difficult task of reconciling the diverse, and even conflicting, requirements of the various "customers" who require the project.

In addition to questioning requirements, a key element of design optimization is *getting all the customers involved in the design process.* For example, the following organization functions might be involved early in the project when requirements are being set, as well as at later stages when the design is reviewed:

Manufacturing
Maintenance
Construction
Safety

The idea is to get their input early, agree on both the design requirements and the design approach, and then have reviews before too much time and effort have been committed.

Design Optimization recognizes that the best opportunities to influence project profitability occur early in the design process. And, the cost of making changes increases exponentially as the project progresses.

Design Optimization—a Definition

We can define Design Optimization as follows: *Design Optimization is the process of defining the optimum design to meet the right set of requirements.* Traditional methods of cost reduction say, in effect: "Here is the design that meets your requirements at the lowest investment cost." Contrast that with Design Optimization, which says: "Here is the design which meets your essential requirements and has the lowest life-cycle cost." The key elements of Design Optimization include:

Life-cycle cost analysis in place of investment cost
Questioning requirements, i.e., seeking to define essential requirements
Creative thinking to generate alternative designs that perform the required
    functions for the lowest life-cycle cost

Constructability—a Key Element of Design Optimization

Constructability is defined by the Construction Industry Institute as "the application of construction knowledge and experience to assure that the design will be time- and cost-effective to construct."

There is much evidence to suggest that constructability studies early in the design process can have a remarkable impact on the time and cost of construction, as well as later savings in operating costs. In a constructability study, a construction expert considers such factors as:

Layout—plans and elevations
Access
Engineering standards that may result in inefficient design
Efficient use of space, manpower and equipment
Local market conditions and construction practices

Elements of Value Engineering

Design Optimization, of course, means implementing some of the principles of Value Engineering. Although often associated with large projects, Value Engineering has much to offer the small project as well.

One of the key principles of Value Engineering is the concept of Functional Analysis. This approach to designing a facility, system or component requires that we think of the *function* of that facility, system, or component. We should ask ourselves, "what does it do?" Functions are expressed in a verb-noun format. For example, the function of a tank is to "store fluid." By focusing on functions, more creative solutions can be found.

Note the differences between Value Engineering and traditional cost-reduction methods. Traditional cost reduction would look at a tank and say: "What is that? It is a tank. How can we design and build it for a lower investment cost?" On the other hand, Value Engineering looks at a tank and says: "What is the function of this facility? It is to store fluid. How can we provide this essential function at lower life-cycle cost?" The answer, developed by creative engineering, might be to use underground caverns for storage.

PROCEDURES FOR DESIGN OPTIMIZATION

Given a specified system or component, design optimization wants to know: "How can its quality be maintained and its life-cycle cost reduced?"

To determine the answer, the design optimization must answer the following questions:

1. *What are the functions of this system or component?* Note that fucntions are expressed in verb-noun format. For example, the primary function of a pump might be "transport fluid."

2. Of the functions, *which are essential and which secondary?* A secondary function of a pump used in manufacturing might be "provide redundancy." This means that two pumps might be installed, each capable of providing the entire load in the event the other one failed. (Note that, if our pump was part of an aircraft hydraulics system, 100% redundancy would probably be considered an essential function. In that case, "minimize weight" might be a secondary function.)

3. *What would be the minimum net-present-value life-cycle cost of providing the essential functions?* This cost can be defined as the "worth" of the system. In our manufacturing example, the minimum cost to provide the required function would be the life-cycle cost of one pump and driver with all associated piping, electrical, instruments, supports and controls.

4. *What is the net present value life-cycle of providing the system or component as it has been specified?* In our example, the cost to provide 100% redundancy would be the cost of two pumps and drivers with associated piping, etc. This cost would probably be about 50% greater than the cost of one pump, as much of the piping and other appurtenances need not be made redundant.

5. *What is the ratio of the cost to provide essential functions versus the cost as specified?* This ratio is an expression of the "value" of our base design. If the ratio is high, the value is low and alternative design approaches may be called for. In our example, the value is 1.5, which is not a bad ratio. The customer and designer might look at the net present value life-cycle cost of both design alternatives (i.e., one pump versus two). If the incremental cost for two pumps can be justified, then two pumps can be provided.

6. *What design alternatives can be developed, which provide essential functions, and maintain quality while reducing life-cycle cost?* These alternatives can be developed through a process of "brainstorming." This creative thinking as actually quite enjoyable and should be done in small groups. Ground rules for a brainstorming session include the acceptance of all ideas (not matter how bizarre) and concentration on the functions to be performed by the system or facility being designed.

Opportunities for Design Optimization

Design Optimization is best performed at the times in the evolution of a project when it can do the most good. Clearly these are:

During the initial scoping phase, when requirements are being set
During the preliminary design phase, when major systems and components are being specified
During the detailed design phase, when specific detailed design decisions are being made

CHAPTER SUMMARY

In this chapter, we discussed the importance of the design phase of a project. The goal of profitability is best achieved when careful attention is given to design optimization. Design optimization requires examination of the requirements, consideration of life-cycle costs, and the use of creative engineering to develop cost-saving alternatives.

11
Management Reporting: The Key to Project Control

PROJECT CONTROL: A MATTER OF PERSPECTIVE

Many project leaders look at a project and see the large number of people performing various tasks, discussing problems, holding meetings, writing memos and reports, making telephone calls, etc. In the field, of course, they see lots of construction equipment, more people, and more problems. Their day is often spent in going through the ever-overflowing in-tray, returning phone calls, attending endless meetings, plowing through voluminous reports, preparing voluminous reports, and just generally trying to keep up with what is going on. Contrary to popular belief, very little control over a project can take place in this environment.

Everyone who has flown in an airplane has noticed how different things look from that lofty perspective. The snarled traffic that we fought through to get to the airport now seems more of a curiosity if we notice it at all. We see our cities and towns in a much broader context from the perspective of the air. Successful project management also requires a different perspective: The ability to rise above the day-to-day rush of activities and get a clear picture of what is really going on. So, how do we get the "airplane" perspective on our projects?

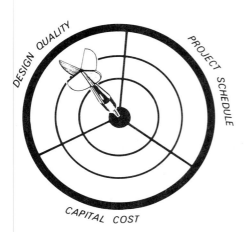

Figure 11.1   Project-management target. This shows that the optimum combination of quality, schedule, and cost is that which produces the maximum profitability.

Effective project control requires only that we know, at all times, the answer to four very simple questions:

If current trends continue, when will the project be complete?
If current trends continue, what will the project actually cost?
What are the causes of adverse trends?
What is being done to correct adverse trends?

These questions can be answered by the collection, analysis, and presentation of project data. Data, properly presented, can give us the airplane view we need to raise our sights from the detailed day-to-day activities to the target of those activities: the optimum blend of cost, time, and quality. Project data gives us the feedback we need to adjust our sights; to define, aim at, and hit the "target" that is the optimum combination of project variables (see Figure 11.1). If our current direction is off-target, we should be able to see why, and what can be done to correct our aim. The way in which data can be used for this purpose is discussed in this chapter.

EFFECTIVE CONTROL REQUIRES EFFECTIVE REPORTING OF PROJECT DATA

Project control takes place when those who are in a management or supervisory role are aware of the need for corrective actions, take the necessary actions, and followup to assure those actions were effective. The individuals who are in a position to effect project control include project managers, project engineers, design supervisors, purchasing agents, construction superintendents, craft supervisors, foremen, and anyone else who has the authority to affect the way people are working, either inside or outside the company. Even the company president can occasionally become involved in a project-control activity, particularly in cases in which resolution of problems with a contractor or supplier must take place at a high level.

The actions these individuals take are generally of a managerial nature, that is, they affect the way other individuals and organizations do their work. These actions may take the form of memos and meetings to discuss specific problems and develop solutions, as well as more drastic steps such as major changes to the contracting or construction plan. But what prompts a project-control action in the first place? The answer is *information*, namely:

What was expected to happen? (the project model)
What did happen? (data representing work to date)
How has actual performance varied from the budget plan?
What trends can be identified from performance to date?
What are the forecasted final cost and completion date, based on performance to date?
What problems can be identified from analysis of performance to date?
What are the possible corrective actions that can avoid delays and overruns?

So it is evident that the quality, quantity, and timeliness of management information is the key to effective project control. We can summarize this concept by saying that project control is simply a matter of presenting:

The right information
In the right format
To the right person
At the right time

This concept is illustrated in Figure 11.2, which is described as follows:
1. A project actually consists of many people doing a lot of different things, with various results. This is the "real-life" situation that we represent with data. For example, progress is measured by data summarizing the total work accomplished and work-hours representing the hours actually spent; the relationship of work-hours spent to work accomplished indicates

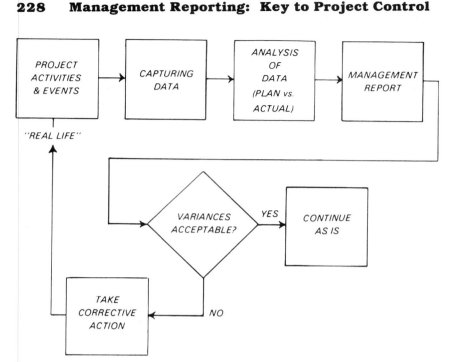

Figure 11.2   Elements of project control.

productivity. So the first step in project control is capturing data that will represent what is really going on.

2. Once we have our data, we subject it to analysis, that is, we arrange it in a certain format and compare it to our original plan to identify variations to date and/or anticipated variations. For example, we might summarize our work-hour expenditure data for comparison with the original project model, and calculate the variation in the actual work-hours expended to date for each labor discipline, against the planned work-hour expenditure.

3. The analysis of data gives us an indication of how the project is progressing. This assessment of the project's status and outlook is summarized in a report. Preparation of this management report usually involves a narrative description of project status, tables of data to pinpoint the facts, and graphical displays to summarize the data. For example, we might note that the work-hours expended in electrical work were greater

than planned, while the physical progress is less than planned. Our narrative would call attention to this fact, point out the cost and schedule impact if nothing is done, and suggest corrective actions that could be taken.

4. After reading the reports, management has a decision to make: are the status and outlook as presented in the report satisfactory? If so, nothing special needs to be done and the project can continue as it was. If, however, the results to date and/or the anticipated final results are less than management expects, some corrective action is required. The report should, in fact, highlight the problem areas and recommend corrective action. For example, corrective action to offset the adverse trends in the electrical discipline might be to review the status of delivery of electrical materials and drawings to the field, the performance of the electrical subcontractor, and the coordination of his work with that of the other crafts.

5. The corrective action is then followed up to assure that the desired impact on the real-life situation is achieved.

So we can see that management reporting, far from being a dull, after-the-fact recitation of figures, is the key to effective control.

TIME: THE MOST CRITICAL ELEMENT

For management reporting to be an effective means of project control, it is essential that the time to complete the "loop" of information flow (as shown in Figure 11.2) be as short as possible. On large projects, it often takes four to eight weeks to progress through the loop, which implies a period of up to 12 weeks from the time something happens to the time something may be done about it. For small projects, which may be completed in only a few weeks, this is clearly unacceptable. In fact, for small projects, speed of reporting is more important than accuracy since, if the information is not available when timely, it will be virtually useless. This means that control of multiple small projects requires a very short cycle for management reporting. To accomplish this short cycle, we will use a combination of shortcut techniques as well as computer-assisted methods.

THE PROJECT MODEL PROVIDES THE BASIS FOR CONTROL

The project model (developed in Chapters 3, 4, and 5) serves as our "road map" for the project-execution phase. In other words, it describes every aspect of how we intend to utilize resources and time to accomplish the given scope of work.

When using a road map to navigate, we compare the map's information on what we should see (towns, roads, etc.) with what we actually see as we drive along. We use this information to calculate where we are, on the map, and compare that to where we planned to be. In this way, we are able to deduce whether or not we need to correct our course, and, if so, how to proceed. We can also predict what will happen if we do not change our course, as well as the time we expect to arrive at our destination, and whether we will be late. The project model is used in exactly the same way. As the project progresses, we monitor our progress relative to the project model, comparing "where we are" in terms of physical progress, time, work-hours, and cost, with where we expected to be. When we seem to be "off course," we can use our project model as a guide as to how to proceed. The difference between the planned and actual values of the project-control variables is the "variance," and it is through the identification and analysis of variances that adverse trends and problems are identified, and control effected.

The project model, representing the design, planning, scheduling, resourcing, and cost basis of the project, can be used to identify variances and to forecast the result if they are allowed to continue. The original model is our fixed reference point, and can be referred to as a "static model." The project as it stands today adjusted for approved changes can be represented by a "current model," reflecting changes to date. Our projection as to the final outcome of the project can be represented by a "forecast model." If these integrated models are computerized, it is relatively simple and quick to update the data and prepare useful reports in which the project models are compared.

DEFINITION OF PROJECT-CONTROL TERMINOLOGY

Before proceeding further with the techniques of project control, it is important to define our terms. Like most other aspects of project management and cost engineering, the terminology is often ill defined, and used inconsistently. The definitions shown below are those that the author has found to work well in practice.

Physical progress: A measure of the amount of work done to date, based solely on physical accomplishments, expressed as a percentage of the current total approved scope of work.

Work-hours spent: The total direct work-hours that have been spent.

Productivity: The ratio of planned work-hours to accomplish a given scope of work to the actual work-hours spent.

Learning curve: The measurable tendency of the time required to perform a given task to decrease as the number of times the task is performed increases, until a maximum efficiency is reached.

Direct work-hours: work-hours that result in measurable physical progress.

Indirect work-hours: work-hours that are required to support the project activities, but that do not contribute to physical progress.

All-in hourly rate: an "all-inclusive" cost per work-hour that includes some overhead and/or indirect as well as direct costs.

Committed cost: The amount of money that should be set aside to cover the forecast final cost of all current purchase orders, contracts, and subcontracts associated with the project.

Cancellation cost: The amount of cost that would be incurred if the project were cancelled and a fair settlement made on all current contracts.

Expenditure: The amount of money that has actually been spent on the project, i.e., the total of all company outlays to cover project charges received to date.

Value of work done: The total cost that would be incurred if all the work done to date were to be paid for according to contract.

Design change: A change, initiated by the design function, which alters the specific way the design is executed. Design changes are experienced on virtually all projects, and provisions should be made for them in the plan and estimate.

Scope change: A change to the basic specification of the project. A scope change adds facilities or capabilities that were not previously part of the project. Scope changes are generally not provided for in the plan and estimate nor covered by contingency. Note that a scope change to a contractor, increasing his assigned scope of work, may not be a scope change to the project.

Field change: A change made in the field to facilitate construction. All projects experience field and startup changes, and these should be allowed for in the plan and budget.

Startup change: A change made in the field to facilitate the startup or operation of the facilities.

Punch-list: A list of small jobs that must be done before the unit is considered complete and ready for startup (also known as a "but-list").

Contingency rundown: A systematic reduction of the contingency included in the cost forecast, to reflect the addition of changes and the reduction of uncertainties.

Cost and schedule forecast: A prediction of the final cost and completion date of the project if present trends continue.

Trend curve: A curve that plots project variables, such as progress and time, and, by using extrapolation and/or a standard curve shape (such as an S curve) can predict the final result.

"S" curve: A standard curve describing project variables over the project's life.

GUIDELINES FOR MANAGEMENT REPORTING FOR MULTIPLE SMALL PROJECTS

Use Management by Exception

On small projects, we can expect that any management attention we receive will be "by exception." In this context, "management by exception" refers to the fact that systematic, in-depth management review of the status of multiple small projects is not likely. What is likely is that management attention will be obtained when their attention can be focused quickly on those items that need it. This is equally true for the project leader, whose attention is often divided between many projects and the various day-to-day problems of each. Management by exception is made possible by reporting formats that highlight problem areas.

Management Reports Must Be Easy to Prepare

Due to the lack of both time and resources, management reports must be easy to prepare. That means that the reports must be free from any information that is nonessential or noninformative. The report should be complete, but just complete enough to advise of general status and convey the information needed when corrective action is required. To be easy to prepare, the report must utilize, wherever possible, input data that is easy to obtain and process.

Computerization of Report Preparation

In order to achieve the rapid processing of data and preparation of reports, computer-assisted techniques are very appropriate in the multiple small-project environment.

One feature on many project managers systems, which can help in accomplishing management by exception, is the ability to sort, select, and present project information in a user-defined format. For example, a program can be set up to produce a report that summarizes all the current projects and lists the projects in decreasing order of overrun amount.

Another way in which computer systems can help shorten the time required for report preparation is by maintaining a current file of relevant information, such that the report preparation requires little more than a printout of the current data.

Many of the current project management systems are designed to produce progress reports, and can be integrated with software for word processing, "spreadsheet" calculations, data storage, and management graphics. More sophisticated systems can customize report formats, and tie the computer directly into other company systems.

TRACKING CURVES: A USEFUL TOOL FOR SMALL PROJECTS

Tracking Curves: A Good Shortcut Approach

A tracking curve is the means by which we can show:

The variations between planned and actual performance
The forecasted final result, if nothing is done to correct current variations

Tracking curves have the additional advantages of being easy to prepare and graphical in their presentation, so that a great deal of useful information is conveyed quickly. They are, therefore, a very useful tool for multiple small projects in which simplicity and time are of utmost importance.

What Is a Tracking Curve?

The purpose of a tracking curve is to represent, graphically, the variation in performance to date from the planned performance. By "tracking" performance to date, the curve also permits extrapolation to forecast the final results if present trends continue. The basic elements of a tracking curve are shown in Figure 11.3, and are described as follows.

The Dependent Variable

The dependent variable, drawn on the y-axis, is the project-control variable whose performance and trends we wish to monitor, forecast, and control. Typical project-control parameters include:

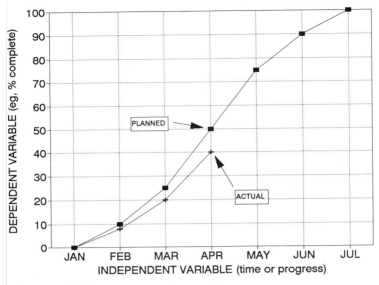

Figure 11.3   Elements of a tracking curve.

Physical progress (i.e., % complete)
Work-hours expended
Cost of work done
Productivity
Unit costs
Hourly costs
Number and cost of changes

The Independent Variable

The independent variable, drawn on the x-axis, describes that project-control parameter that we wish to use as a measure of "where the project is." The two most common independent variables are:

Time (i.e., days, weeks, months, or dates)
Progress (i.e., physical % complete)

The Reference Curve

The reference curve, often referred to as the S curve, describes the relationship between the dependent and independent variables that has been anticipated by our project model. It might show:

The rate at which we expect to make progress (i.e., the % complete per month)

The rate at which we expect to expend work-hours or costs

The reference curve often takes the form of an "S" on large projects, because large projects often experience first a period of accelerating progress, then a steady-state rate of progress, and finally a decelerated rate of progress. This is illustrated by Figure 11.4, which shows the stages of construction on a large project, and is described as follows.

1. In the *acceleration stage*, civil work predominates. Site clearance, excavation for foundations and underground lines, installation of temporary construction facilities, and construction of foundations all occur during this phase. The rate of progress is often limited, because the other craft operations (mechanical and electrical work, ironwork, etc.) cannot begin until the preliminary civilwork is almost complete, and there is a limit on just how quickly that work can be progressed. The number of people on the job usually builds up during this stage.

2. In the *steady-state stage*, all crafts are able to work on the job. The number of people reaches a peak during this stage, and remains rela-

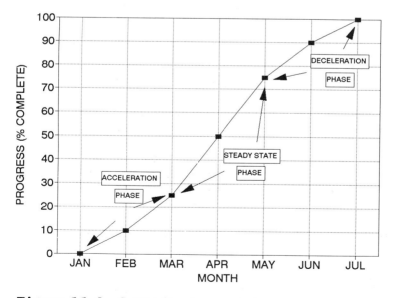

Figure 11.4 S curve for a large project.

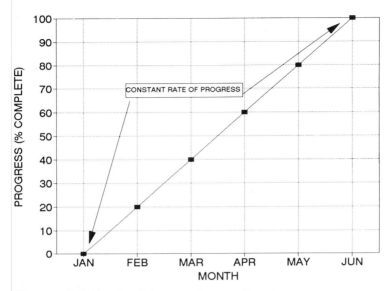

Figure 11.5   Possible curve for a small project.

tively constant. The rate of progress also reaches a peak and remains relatively constant as well.

3. In the *deceleration stage*, the work of the major crafts is essentially complete, and the number of people is reduced accordingly. The rate at which progress is made also, of course, decreases. The type of work done during this stage (painting, insulation, startup changes, "punch list" items, etc.) also does not appear to contribute much to measurable physical progress.

But what of the small project? What shape does its reference curve take? For a small project, it is important to recognize that *the reference curve will not necessarily be an S curve.* The small project may not require all the various crafts or have a period of manpower buildup or rundown. In fact, many small projects, such as turnarounds or other maintenance operations, have a constant number of people from start to finish (see Figure 11.5).

Figure 11.6   (a) Cumulative progress curve for a small project. (b) Progress curve in histogram format.

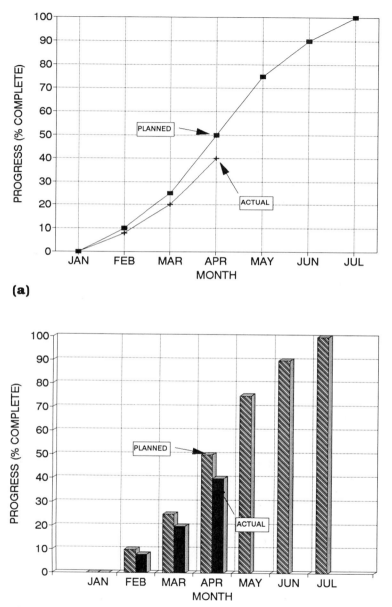

(a)

(b)

Tracking-Curve Format

There are several different formats available for tracking curves, the selection of which depends on the type of variables being used, the preference of the user, and any computer hardware or software limitations that might apply. The different configurations are illustrated in Figure 11.6, and are described as follows.

Cumulative   This illustrates the cumulative (i.e., up to-date) value of the dependent variable (see Figure 11.6). For example, the dependent variable might be "percent complete earned to-date." This type of curve is usually continuous.

Incremental   This illustrates the incremental (i.e., for-the-period) value of the dependent variable. For example, the dependent variable might be "percent complete earned this week." This type of format is usually shown as a histogram.

Combined   This kind of tracking curve combines both the cumulative and incremental types on one curve. For those who are used to seeing such curves, this format can be quite useful. However, most other people find it somewhat confusing.

Differential   This tracks the deviation of the dependent variable from a reference value (see Figure 11.8b). The deviation is usually considered to be acceptable if it is within predetermined limits. Trends that indicate that the final value will be outside the limits are a signal that a problem is developing.

Setting Up Tracking Curves

Step 1: Select the Dependent Variables to be Controlled

Each company, and often each project, has certain project-control variables that are particularly important. Typical variables are shown in "The Dependent Variable," earlier in this chapter.

Step 2: Select the Independent Variables

Typical variables are shown in "The Independent Variable," earlier in this chapter.

Step 3: Select the Format to Be Used

See "What is a Tracking Curve?" earlier in this chapter.

Step 4: Draw the Reference Curve

It is evident that the successful use of a tracking curve depends on the reference curve's being well thought-out and realistic. If it is not, the variations between planned and actual values will be useless as a tool for forecasting or control. Many users of tracking curves draw the reference curve by using judgements or standard curve shapes (such as the "S"). For small projects, this is unsatisfactory.

The project model described in "The Project Model Provides the Basis for Control," provides the basis of the reference curve. Because it defines the work to be done, the resources required, and the direct work-hours to be expended in each segment of time, it can be used to derive the reference curves for direct work-hours vs. time, physical progress vs. time, total work-hours vs. time, expenditures vs. time, etc. The progress vs. time curve can then be used to plot other variables against progress.

In deriving the reference curve of progress vs. time from the project model, we can assume that the productivity is constant with time, or that it varies in a predetermined way. Many companies involved in large projects use a productivity trend curve that assumes lower-than-average productivity during the acceleration and deceleration stages, due to "learning curve" and other effects. However, since small projects, in general, do not have a significant acceleration or deceleration stage, the planned productivity can generally be assumed to be constant. Physical progress will then be achieved over time at the same rate at which the direct work-hours are expended.

As described in "Fundamentals of CPM," in Chapter 3, noncritical activities in the project can be scheduled as if they will start on their early-start date, their late-start date, or at some time in between. Whatever assumption is made will, of course, affect the planned progress curve: if all activities start on their early-start date, progress will be made more quickly than if they start later. Interestingly, although it is rare for all activities on a project to start on the early-start date, many projects are scheduled as if they will. As a result, progress then seems to lag behind schedule. It is very useful, therefore, to construct progress curves in which the progress for both early-start and late-start assumptions are shown. The actual progress will, hopefully, fall between these two extremes. An example of such a curve is provided in Figure 11.7 which shows both early-start and late-start schedule.

Sample Tracking Curves

Figure 11.8 shows some sample tracking curves of the cumulative, incremented, and differential type.

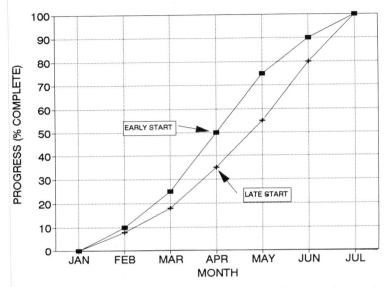

Figure 11.7   Cumulative progress curve showing start-date variations.

Using the Tracking Curve for Forecasting and Control

Identifying Variances in Work to Date

In Figure 11.2, we see that the key to effective control lies, to a great extent, in the process of data analysis. Given that we have captured data that represents what is actually happening on the project, we must now use that data to figure out:

What has happened up to now
What is going to happen if things continue as they are now
What problems might be causing things to go wrong
What corrective action might be taken to avoid or correct those problems

Figure 11.8   (a) Cumulative work-hours as a function of physical progress. (b) Direct work-hours spent per % of physical progress. (c) Direct work-hours vs. planned work-hours vs. earned work-hours.

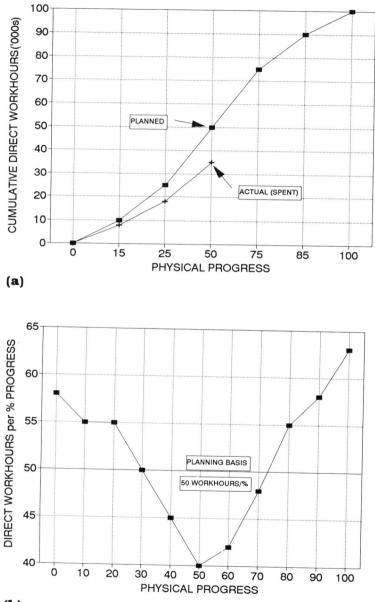

(a)

(b)

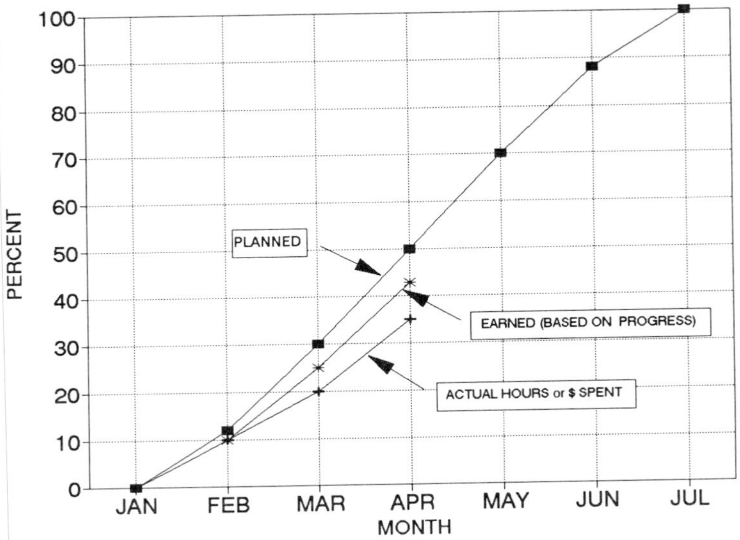

Figure 11.8c

This information is then presented in management reports, with the intention that management (that is, whoever has the appropriate authority) take the necessary action.

   To answer these questions using tracking curves, it is often helpful to look at two or more tracking curves together. For example:

If: Progress: is below planned progress to date
     Work-hours: are below planned work-hours to date

We might conclude that inadequate manpower resulted in lack of progress

If: Progress: is below planned progress to date
     Work-hours: are per plan

We might conclude that productivity is below expectations

In fact, there are many combinations of project-control variations, and each one can tell a very different story. In analyzing project data, the project engineer often has to be a detective, using the data as "clues" that when taken together, form a picture of what is really going on.

Checklists are a good way to find our way through the maze of project data. We can begin by using a checklist such as that shown on Figure 11.9. The checklist shows that we should generally look at the project-control variables, in the sequence shown, ask whether each control parameter is the same as, greater than, or less than planned. The parameters are as follows:

Progress (i.e., physical % complete): Is progress ahead of schedule, on schedule, or behind schedule?

Work-hours (i.e., work-hours spent to date to achieve the given progress): Are the work-hours spent above, equal to, or below the planned work-hours?

Scope of work (i.e., quantities done to date and left to do): Is the quantity of work done to date greater-than, equal to, or less than the planned scope of work? Is the total quantity of work in the project greater than, equal to, or less than that which was planned?

Productivity (i.e., ratio of planned to actual work-hours): Is productivity greater than, equal to, or less than planned?

Resource availability (e.g., materials, labor, equipment): Has labor been supplied to the job in the quantities and disciplines as planned? Has material been provided on schedule? Have approved drawings been provided on schedule? Has the necessary equipment been provided as planned? Have overhead staff and facilities been provided as planned?

Schedule milestones: Have major schedule milestones to date been met?

CONTROL VARIABLE		AHEAD	ON TARGET	BEHIND
PHYSICAL PROGRESS				
DIRECT HOURS SPENT				
SCOPE OF WORK				
PRODUCTIVITY				
AVAILABILITY OF RESOURCES:				
	MATERIAL			
	LABOR			
	EQUIPMENT			
	FACILITIES			
	DRAWINGS			
MILESTONES ACHIEVED				
EXTERNAL FACTOR IMPACT				

Figure 11.9   Project-control checklist.

External Factors (e.g., weather, strikes): Have external factors had any influence on the project to date? Are external factors expected to influence the work left to do?

To determine the current status of our project, then, we need to look at the answer to each question individually, and also all the questions together. The tracking curves should be designed to help a project leader complete that process quickly, by helping to identify patterns of information.

Forecasting Future Performance

Cost and schedule forecasting is one area of project control in which the philosophy and judgement of the individuals and organizations involved can have a major impact. Some companies and individuals prefer to forecast that the final cost and schedule will be identical to the budget plan, in the hope that any delays or overruns that are currently foreseen will be offset by future lucky breaks. This tendency to cover-up potential overruns or delays has the effect of reducing project-control effectiveness since problems requiring corrective action are not highlighted.

There is also a natural tendency to avoid criticism and punishment. The dilemma of project leaders is often that if they forecast a delay or overrun, there is sure to be trouble right away, but if they report that the project is on schedule and budget, they might (or might not) get into trouble later, if and when the overrun or delay actually occurs.

If a cost or schedule forecast is to be effective as a project control device, it should be defined as follows:

"This cost (or schedule) forecast represents the final actual cost (or schedule) that will be experienced *if no action is taken to correct the problems or adverse trends identified to date.*"

In other words, the forecast is not what we think will actually happen, it is what we think will actually happen if no corrections are made. Therefore, a tracking curve can be used to extrapolate data trends to forecast what the curve of actual data will look like. The extrapolation is usually based on:

1. The variations defined by the plot of actual data to date (e.g., progress per month is 75% of the planned progress rate)
2. The shape of the reference curve (e.g., an S curve)
3. The slope of the trend curve through the actual data (e.g., a "flatter" slope)

Tracking curves are a useful way to assure that the possibility of delays and/or cost overruns cannot be easily hidden. The data points reflecting actual performance will generally follow one of several patterns, as described in the following:

No significant deviation from the plan   In this case, we have good grounds for forecasting that the dependent variable will follow the reference curve and that the project will finish on time and within the budget.

Clear adverse trend   When actual data points show a clear adverse trend, such as progress being consistently below expectations, it is a clear signal that a delay and/or overrun can be expected if nothing is done. We can expect that the actual values of the dependent variable will follow the shape of the reference curve, but that the slope will reflect the trend established to date. So, for example, if we are a week behind schedule because we are making less progress than planned each week, the cumulative effect is that we will be much more than a week late at the end of the job.

Clear favorable trend   If the actual data indicates a clear positive trend, it may indicate that a cost underrun or schedule improvement is possible. However, such favorable results should be reported only at such time as the reason for the favorable results to date can be identified, and it can be expected that the favorable trend will continue.

Scattered data with no clear trend   Tracking-curve data often shows a degree of scatter that is clearly beyond the expected variation. Not only does such data make it impossible to predict trends and make forecasts, but it indicates a lack of credibility in the project-control methods and data. If such data is experienced, it almost certainly indicates that something is wrong with the data collection and analysis method. One frequent problem area is in the definition of terms. For example, it may be that indirect work-hours are being reported as direct, or the progress-measurement system may contain some anomalies. In any case, the problem should be identified and rectified.

It should be noted that some project-control variables, such as productivity, may well experience a scatter pattern, and not exhibit a clear trend.

THE HIERARCHY OF REPORTING MULTIPLE PROJECTS

For management attention to be focused effectively on those areas requiring action, it is essential that reports be concise and to the point. As

PROJECT		BUDGET	FORECAST	VARIANCE	SCHED.	FORECAST
NO.	NAME	($,000)	($,000)	(%)	COMPL.	COMPL.
123	PUMP SERVICE	25500	21000	-18	MAY 5	MAY 5
132	CLEAN VESSELS	57000	55000	-4	JULY 16	JULY 16
143	ROAD REPAIR	75000	85000	13	SEPT 3	SEPT 10
145	REROUTE PIPING	155000	160000	3	MAY 9	MAY 10
153	NEW SWITCHGEAR	74000	85000	15	AUG 2	AUG 16
177	NEW ROOF ON LAB	56000	55000	-2	AUG 10	SEPT 6
187	REPLACE VALVES	124000	129000	4	DEC 5	DEC 10
	TOTAL	566500	590000	4		

Figure 11.10   Small-project summary report.

shown in Figure 2.5, the level of detail required by management is a function of the level of the user in the organization: the higher up in the organization the user is, the less detail is important.

It is possible, therefore, to visualize a family of reports, all generated from the same project model, in which the level of detail is matched to the level in the organization of the person receiving it. Figures 11.10 through 11.15 show examples of this hierarchy of reporting for progress and cost control. The first report (Figure 11.10) shows only summary information. It is intended for identification of those projects that are experiencing or might experience delays and/or overruns. This report might be of interest to the manager who is responsible for the department that handles all the small projects. The second-level schedule report (Figure 11.11) provides additional planning information such as the start dates and work-hours of each project. It is still, however, a summary report on which a number of projects can be displayed. This level of detail might be appropriate for a supervisor of several project engineers, each of whom

PROJECT		COMPLETION		VAR.	PROG. TO DATE (%)		
NO.	NAME	SCHED.	F'CAST	DAYS	PLAN	ACT.	VAR.
123	PUMP SERVICE	MAY 5	MAY 5	0	85	90	5
132	CLEAN VESSELS	JULY 16	JULY 16	0	95	95	0
143	ROAD REPAIR	SEPT 3	SEPT 10	7	33	30	-3
145	REROUTE PIPING	MAY 9	MAY 10	1	73	65	-8
153	NEW SWITCHGEAR	AUG 2	AUG 16	14	55	50	-5
177	NEW ROOF ON LAB	AUG 10	SEPT 6	27	45	55	10
187	REPLACE VALVES	DEC 5	DEC 10	5	5	10	5

Figure 11.11   Small-project schedule-summary report.

handles several projects. Similarly, the second-level cost report (Figure 11.12) would be used by the same supervisor: it also provides summary information on a number of projects. Note that the "variance" columns permit management by exception. By scanning those columns, the reader's attention is immediately focused on the problem areas.

The third level of report is for the project leader. The purpose of these reports is to pinpoint those specific areas of a given project that are causing the overrun or delay. This report would, in general, be prepared for one project at a time. It shows the kind of detailed information that enables the project leader to identify the specific activities that have problems and, by deduction, the nature of the problem. All project management programs provide a wide assortment of detailed reports for costs, schedules, and assignments.

The project model is most helpful in preparing these reports. If we think of the project model as a hierarchy of information, the lowest level of the hierarchy shows the maximum amount of detail. That level is the level at which the work is actually done. For example, if company workforces are performing the construction work, the management level at the "workface" is the foreman. Below him are the people whose workhours are considered direct and whose efforts result in measurable physical progress. To the foreman, the project consists of those activities for which he is responsible. The resources are the people who work for him, the materials they install, the engineering drawings he needs to do the work, and the construction equipment he uses. His work is planned and recorded using timesheets, tasksheets, job-cards, work-orders, or some other detailed description of specific work scope. Progress is measured at this level of detail, as are manhours. This level of detail is the level of detail of the project model.

PROJECT		BUDGET ESTIMATE	APPROVED CHANGES	CURRENT ESTIMATE	COST FORECAST	VARIANCE (%)
NO.	NAME					
123	PUMP SERVICE	25,500	500	26,000	21,000	-19.23
132	CLEAN VESSELS	57,000	1,000	58,000	55,000	-5.17
143	ROAD REPAIR	75,000	(2,500)	72,500	85,000	17.24
145	REROUTE PIPING	155,000	14,000	169,000	160,000	-5.33
153	NEW SWITCHGEAR	74,000	4,000	78,000	85,000	8.97
177	NEW ROOF ON LAB	56,000	(2,900)	53,100	55,000	3.58
187	REPLACE VALVES	124,000	1,500	125,500	129,000	2.79
	TOTAL	566,500	15,600	582,100	590,000	1.36

Figure 11.12 Small-project cost-summary report.

At the next level in the small-project management hierarchy, we might find the project leader (or the field superintendent, construction manager, site engineer, etc.). Regardless of his or her title, the person at this level has a perspective that encompasses the entire project's scope and schedule. The information required to manage at this level can be derived easily by summarizing information from the detailed level.

The next level of detail for the small project is generally one in which several projects are considered at once. These reports can also be prepared by aggregation from the project models. For those using the higher-level reports with summary-type information, the additional detail available at the lower level reports is used for problem identification, analysis, and resolution. It can be thought of as an "audit trail": there when needed but not evident when not needed.

The essential points about the hierarchy of reporting are these:

1. The perspective that is essential to effective project control can only be attained when excessive detail is eliminated from management reports.
2. The level of detail should match the level of the person using the data.
3. Reporting should facilitate management by exception.
4. Detailed information should be available and used for problem identification and resolution.

USING REPORTS TO MANAGE THE WORK OF OTHERS

The Problem of Objective Reporting

The classic problem of project control stems from the following situation:

The project leader (or manager) is not in a good position to collect and analyze detailed information, so therefore he must rely on the contractor to do it.

The contractor (or subcontractor) cannot be blamed for having a natural aversion to reporting "bad news," particularly if it reflects poorly on his own performance.

To put it another, more general, way, progress reports are intended to report on the performance of the people doing the work to those who pay for it. However, the only people who are in a position to collect data and prepare the progress reports are those whose performance is being measured and reported. It is only natural that, under those circumstances, there is apt to be a certain lack of objectivity in reporting, particularly when those preparing the report are certain that, if the report contains unfavorable information, a lot of trouble will result. It is easy, under those circumstances, to convince oneself that there is no need to report adverse trends because something good may happen soon that will offset those cost overruns and schedule delays.

Yet, there can be no question that timely, objective reporting is essential to project control. How then can we assure that the natural bias against reporting accurately is overcome? The solution is to design and implement a project-control system that leaves no room for "cheating." This requires that the basic project data, such as work-hours and physical progress, be captured, analyzed, and reported in a rigorous way. Computerized systems, of course, are an aid in this. Usually, one finds that the data required for payment of an invoice, such as work-hours or progress, is carefully monitored and reported and therefore forms a good basis for control.

Using the Hierarchy of Networks

If we can break a project down into its component packages of work, and then manage and control those packages, we generally find that the smaller a work package is, the easier it is to control. Therefore, the use of "hammocks" and "subnetworks," as described in "Definition of Planning Terminology," in Chapter 3, (see Figures 3.3 and 3.4), enables the project leader to measure progress and work-hours at a lower level of detail than he would normally work in. For example, progress measured at the subcontract level can be aggregated to reflect progress in the major subprojects that can, in turn, be aggregated to reflect progress in the entire project.

Progress measurement methods are discussed in Chapter 12. On small projects, where formal progress measurement systems may not be available, judgement is often the only way to measure progress. This method, discussed more fully in Chapter 12, requires that the work packages on which progress is to be measured be small enough that the judgements are likely to be reasonable. The larger the work package, the more difficult it is to measure progress by judgement.

CAPTURING DATA

In the operating-plant environment in which many small projects take place, it is unlikely that project leaders will be in a position to force substantial changes to existing practices and procedures, so that they can have the project control data they need. It is important, then, to carefully identify data that is available from existing procedures. Project data can then be selectively captured by tapping into the flow of data that already exists.

Before project-control methods, systems, and procedures are designed, a careful review should be made of all existing practices and procedures. The specific documents that provide project data such as man-hours, costs, resource status, etc., should be identified. The data on each document that is required by the project leader should then be identified. This data should be kept to a minimum to allow it to be captured quickly. The means by which the project engineer will store the data and use it to update the project model is then defined: it might be by manual or computerized input, or even perhaps by a transfer of computer files. Finally, the data needed for project control that is not available from existing sources is identified, and procedures established for capturing it.

CHAPTER SUMMARY

In this chapter we discussed the importance of management reporting to the project-control function. We described the use of tracking curves as an appropriate forecasting and control technique for small projects, the use of the project model as a basis for control, and hierarchy of plans that can be built around the project model to provide timely information at each level in the organization.

12

Measuring Progress and Performance

THE IMPORTANCE OF
PROGRESS MEASUREMENT

Project control depends, to a very great extent, on the effectiveness of the methods used to measure and report progress and performance. Even on small projects, a progress-measurement method is essential as it defines "where we are" such that we can compare that with "where we planned to be," and thereby identify areas needing attention.

Small projects, because of their size and other aspects of the small-project environment, often experience ineffective project control because of the lack of progress measurement. Proper progress measurement is often thought to be simply too much time and effort for the small project, but this need not be the case. A progress-measurement method can be defined once, and then implemented uniformly on all small projects. In this chapter we explore the basic principles of the "Earned Value" method of progress measurement, and then discuss various shortcut techniques that can be applied.

PRINCIPLES OF PROGRESS MEASUREMENT

Elements of an Effective Progress-Measurement Method

An effective progress-measurement method should do or be the following:

Provide a measure of the physical quantities of work done
Provide a measure of the current total scope of work of the project
Express the work done as a percentage of the total current scope of work
Be unbiased (i.e., it should not be significantly affected by optimism or pessimism)
Be realistic (i.e., it should reflect the many hard-to-measure items of work that, individually, may be small but which, collectively, contribute significantly to the scope of work)
Be agreed upon (by those whose performance is being measured and by those who are doing the measuring)
Be fair (to those who are doing the work and to those who are paying for it)
Be efficient (i.e., it should not require excessive time to collect, analyze, and present data)
Be well documented (to assure consistency)
Be independent of actual work-hours and costs (i.e., it should measure actual physical work done without regard to the number of work-hours spent)

Basic Steps in Progress Measurement

The basic steps that are required to set up and implement a progress-measurement method are described below. Each step will be illustrated in "The Earned Value Method of Progress Measurement," later in this chapter.

Step 1: Divide Scope of Work into Packages for Control

Most project-management techniques are based on breaking a project down into components or packages of work, the idea being that each package of work be well-defined and appropriately sized to the level of the person or company handling it. Often a "work package" is assigned to a specific individual, organization, function, or contractor. Examples of work packages are:

An activity on a network plan
A work package in a "work breakdown structure"

A contract or subcontract
A "work order," "task-sheet," or "job-card"
A variation to a contract
The work of one craft or discipline
A specified quantity of work
The work in a given geographical area
The work covered by a certain code in the code of accounts

For progress measurement, it doesn't really matter what form the work packages take, so long as *the total amount of work in all packages equals the total scope of work of the project.* Naturally, it is preferable that progress be measured against network activities so that the progress to date can be used as the basis for a schedule forecast, but this is not essential. Note that activities should be "definable, assignable and significant" in order to facilitate effective progress measurement (see Chapter 3).

The scope of work may actually be divided several times until sufficient detail is reached. The work packages should also be compatible with the way work-hours are collected, to facilitate productivity measurements.

Step 2: Establish the Standard Work Unit

If progress in diverse activities is to be measured and then aggregated, it is helpful to have a unit of measure that can be applied to any activity, regardless of the type of work being done. For example, we could not add progress in the form of cubic yards of concrete poured to progress in the form of linear feet of electrical cable installed, yet we would like to be able to calculate the net progress for both civil and electrical work. To do so, we define a standard work unit with which all progress can be expressed. The most frequently used standard work unit is the "planned work-hour." This is the basis of the "earned-value" system of progress measurement that will be described later in this chapter. The advantage of the planned work-hour approach is that earned work-hours (based on physical progress) can be easily compared with actual work-hours to give an indication of productivity. In many cases, costs are used instead of work-hours.

Step 3: Define the "Yardstick" for Measuring Progress

This step consists of defining, *in advance*, what amount of work or achievement of milestones constitutes what % complete. This can be done by judgement, experience, or by an allocation of estimated work-hours. It consists of identifying the physical results of the direct labor in the project, as well as the milestones or steps required to complete the work on each item. The yardstick is the key to the "earned value" method.

Step 4: Define the Method for Aggregating Progress

This step consists of defining the means by which progress in one category will be added to progress in other categories to arrive at the overall % complete.

Step 5: Define Who Will Measure Progress, and How Often

The individuals or functions who will actually do the progress measurement should be defined as part of the procedure. It should be clearly understood who will do the measuring, what data will be required, how often it will be required, who will collect and analyze the data, to whom it will be presented, etc.

Step 6: Agree on the Progress-Measurement Method
Prior to the Start of Work

Progress measurement is a combination of numerical analysis and judgement. Because the individual or organization whose performance is being measured must, in general, provide much of the information necessary to measure performance, it is evident that good cooperation and communication is essential if the system is to work. For this reason, it is important that the method for measuring progress be agreed upon prior to the start of work. Both those being measured and those doing the measuring must agree that the method for measuring progress is fair.

ESTABLISHING A PROGRESS-MEASUREMENT METHOD FOR SMALL PROJECTS

Although small projects often seem to be too small to justify a formal progress-measurement system, the fact is that, once a progress-measurement system is established it can be used over and over again on all small projects. Just as in planning, contracting, cost estimating, and other project-management functions, the benefits of increased consistency, effectiveness, and efficiency far outweigh the effort required to develop and implement such methods.

The progress-measurement system should cover the three major functions of a project: design, procurement, and construction. (Of course, not all projects contain all three functions.) The progress-measurement system should cover all work that is considered direct, and should focus, whenever possible, on the *physical results* of the work, instead of the effort required. For example, it is often said that it is difficult to measure progress on work done by engineers, scientists, programmers, and other "white-collar" persons. In such cases, we can ask ourselves, "What physical result is produced by the efforts of this person?" Perhaps it is a drawing, report, computer program, design specification, bill of materials, purchase order, contract, etc. When the items that represent the physical work are identified, it is a simple matter to identify the major milestones involved and create a progress measurement yardstick by assigning %-complete values to them.

THE EARNED VALUE METHOD OF PROGRESS MEASUREMENT

The best-known method of progress measurement is the earned value method in which the work done is expressed in terms of work units earned. In most cases, the work units are planned direct work-hours (also referred to as base work-hours, "norm" work-hours, or estimated work-hours). As progress is made and measured using the yardstick, we "earn" planned work-hours, and can sum these earned work-hours to yield a calculation of overall progress. (Note that the cost of the work planned and completed can also be used, in lieu of workhours, as the common work unit.) The earned-value method also provides a good illustration of the basic approach that will be used in all progress-measurement methods. Each step of the earned value method is illustrated in the following example.

This construction project involves subcontracts for civil, architectural, mechanical, electrical, and steelwork.

Step 1: Divide the scope of work into packages for control

Let the major packages be the major subcontracts:

   a. Civil
   b. Architectural
   c. Mechanical
   d. Electrical
   e. Steelwork

Let the component packages be network activities as follows:

   a. Civil subcontract
      1. Site clearance and grading
      2. Foundations
      3. Paving

   b. Architectural subcontract
     1. Walls
     2. Roof
     3. Glass
     4. Insulation, paint, etc.
   c. Mechanical subcontract
     1. HVAC (heating, ventilating, and air conditioning)
     2. Pumps and piping
   d. Electrical subcontract
     1. Transformer
     2. Switchgear
     3. Cable and conduit
     4. Environmental control system
   e. Steel erection subcontract
     1. Heavy steel
     2. Light steel

Step 2: Define the standard work unit

The standard work unit will be planned, direct work-hours (work-hours at the productivity assumed for the plan and estimate). These are taken from the current model, i.e., the project model adjusted for approved changes.

Step 3: Define the yardstick for measuring progress.

For example:

Planned activity	Step or milestone	Percent earned when complete
a. Civil subcontract		
1. Site clearance and grading		
	grub and clear	25
	cut and fill	35
	rough grade	20
	final grade	20
		100
2. Foundations		
	excavation	25
	forms	15
	reinforcing bars	20

		25
pour and cure		25
strip forms and backfill		15
		100

3. Paving

grading	45
base course and compaction	30
final course	25
	100

b. Architectural subcontract*
c. Mechanical subcontract*
d. Electrical subcontract*
e. Steel erection subcontract*

---

*The yardsticks for the other subcontracts are defined in exactly the same way. To establish the percent earned at the completion of each work package, we could have used judgement or we could have estimated the relative work-hours to be spent on each work package and set the percent earned equal to the percent of work-hours.

Step 4: Define the method for aggregating progress

Aggregating progress at the subcontract level

Subcontract	Component	Percent complete	Planned, direct work-hours	Earned work-hours
a. Civil				
	1. site clearance and grading	23	1500	345
	2. foundations	11	2450	270
	3. paving	5	900	45
Total work-hours, this subcontract:			4850	660

Net % complete = earned work-hours/planned, direct work-hours
= 660/4850 = 13.6%.

Progress on each subcontract is calculated in the same way.

Note that, if the actual hours spent on this subcontract to date was 1320, we could easily measure productivity as the ratio of earned to spent hours, or 660/1320 = 50%.

Aggregating progress at the project level:

Subcontract	Total planned direct work-hours	Percent of total	Subcontract complete (%)	Contribution to project progress (%)
a. Civil	4850	40.8	13.6	5.6
b. Architectural	2650	22.3	8.2	1.8
c. Mechanical	1500	12.6	5.0	0.6
d. Electrical	1650	13.8	7.0	1.0
e. Steel erection	1250	10.5	12.5	1.3
	11900	100		10.3

The construction work is 10.3% complete overall.

Step 5: Define who will measure progress and how often

Progress will be measured weekly by each subcontractor using yardsticks agreed upon with the prime contractor and the client. Progress measurement will be reviewed by the prime contractor and the client.

Step 6: Agree on the progress-measurement method prior to the start of work

The procedures for measuring progress and the responsibilities of each partly are defined in an attachment to the basic contracts. These procedures should be thoroughly discussed and agreed upon prior to the start of work. Once agreed, these yardsticks can be used for all small projects at this location.

MEASUREMENT OF PROGRESS IN ENGINEERING WORK

A common misconception with regard to a professional's activities, such as design engineering, drafting, programming, research, etc., is that its creative nature precludes effective progress measurement and control. This argument is, naturally, often put forward by those whose progress is to be measured. In fact, progress measurement on such work activities can be conducted in the same way that it is on construction.

In general, work that is worth measuring will result in a physical result. For example:

Engineering work results in equipment specifications, bills of material, design reports, process flow diagrams, etc.

Design and drafting work results in drawings (layouts, general arrangements, piping and instrument drawings, electrical one-line drawings, etc.)

Inspection and testing work results in a "certificate of fitness" or equivalent document

Reserach or clinical testing results in reports

Programming results in lines of code or the production of a specified software feature

If we follow the sequence of activities described in "Establishing a Progress-Measurement Method for Small Projects" (earlier in this chapter), we might find that a progress-measurement system for engineering would look something like the following.

Step 1: Divide the scope of work into packages for control

Let the major packages be the engineering departments:

 a. Civil and structural
 b. Mechanical
 c. Process
 d. Electrical

Let the component packages be as follows:

 a. Civil and structural department
  1. general arrangements, plot plans, layout, and excavation drawings
  2. heavy structural steel arrangement drawings
  3. excavation drawings
  4. underground piping and electrical drawings
  5. foundation design analyses
  6. structural design analyses
  7. civil and structural detail drawings
 b. Mechanical department
  1. HVAC equipment design analysis and specification
  2. HVAC layout
  3. pressure-vessel design analysis and specification
  4. rotating-equipment design analysis and specification
  5. fired-heaters design analysis and specification
  6. heat-exchanger design analysis and specification

   c. Process department
     1. Process analysis
     2. heat and material balance calculations
     3. process-flow diagrams
     4. piping and instrumentation diagrams
   d. Electrical department
     1. load analysis
     2. equipment sizing and specification
     3. one-line drawings
     4. electrical detail drawings

Step 2: Define the standard work unit

The standard work unit will be planned direct work-hours from the current model.

Step 3: Define the yardstick for measuring progress

As in the case of construction work, detailed work-hour estimates for each work package can be used as guidelines for assigning progress. Judgement, based on milestones and completion of key activities, can also be used. For example:

Preparation of design drawings

Work package	Percent earned when complete
Preliminary sketches	10
Initial layout and discipline review	25
Incorporate vendor drawings	10
Interdiscipline review and revisions	5
Design complete approved by client	25
Changes and revisions	20
Approval for construction	5

Another example involves design analysis:

Work package	Percent earned when complete
Kickoff meetings and job scoping	5
Initial hand calculations	10
Preliminary computer analysis	20

Work package	Percent earned when complete
Review of preliminary design	5
Secondary computer analysis	20
Interim review and revisions	15
Final analysis	10
Preparation of specification	10
Vendor followup	5

Step 4A: Aggregating progress at department level

Discipline	Activity	Percent complete	Planned work-hours	Earned work-hours
a. Civil/structural				
	layouts	65	360	234
	foundation drawings	75	200	150
	structural drawings	40	120	48
	underground draw-			
	ings	45	140	63
	foundation design	65	220	143
	steel design	75	185	139
	detail drawings	35	110	39
Total work-hours, this department			1335	816

net % complete = earned work-hours/standard work-hours
       = 816/1335 = 61%.

Progress in each department is calculated in exactly the same way.

Step 4B: Aggregating progress for all engineering work

Department	Total planned standard work-hours	Percent of total	Department % complete	Percent contribution to project progress
Civil/structural	1335	38	61	23
Mechanical	750	22	42	9
Process	550	16	50	8
Electrical	850	24	32	8
	3485	100		48

The engineering work is 48% complete overall.

Step 5: Define who will measure progress and how often.

Progress will be measured weekly by each department head and will be reviewed with the project engineer as well as the engineering-division manager.

MEASUREMENT OF PROGRESS IN PROCUREMENT

Like engineering, procurement is a "home-office" activity whose results do not seem to lend themselves well to progress measurement. However, procurement activities are often on the critical path, as the timely delivery of materials and equipment are critical to the schedule on a small project. Therefore, the ability to manage progress in the procurement function can be of vital importance to any project.

Like engineering, procurement work results in the production of identifiable work packages and the completion of identifiable tasks. For example:

Preparation, approval, and award of purchase orders
Traffic (i.e., shipping arrangements)
Expediting
Inspection
Liaison with engineering, planning, warehousing, construction, and other
    groups

Each of these activities can be broken down into discrete tasks, and %-complete values assigned just as in the above examples for construction and engineering work. For example:

Major work package is the award of purchase order for major equipment. *The component packages for this work are as follows (Step 1):*

1. Define general requirements (e.g., a 400 hp centrifugal compressor)
2. Screen potential bidders to identify those willing and able to bid, and establish bidders list
3. Prepare and release invitation to bid documents
4. Receive, open, and evaluate bids and prepare bid tabulation
5. Obtain authorization for purchase order
6. Conclude final negotiations and award purchase order
7. Administer revisions and payments
8. Close out purchase order

A progress measurement yardstick might then look as follows (Step 3):

Measuring progress for each purchase order

Work package	Percent earned when complete
1. Define general requirements	5
2. Establish bid list	10
3. Release invitation to bid	10
4. Bid tabulation	20
5. Authorization obtained	10
6. Award purchase order	15
7. Administer	20
8. Close out	10

Aggregating progress for all purchase orders (Step 4A):

Purchase order	Planned work-hours	Percent complete	Earned work-hours
A101	120	50%	60
A102	150	60%	90
C402	130	40%	52
D503	140	45%	63
total work-hours	540		265

net % complete = earned work-hours/standard work-hours
= 265/540 = 49%.

Aggregating progress for the procurement function (Step 4B):

Work package	Total planned standard work-hours	Percent of total	Package % complete	Percent contribution to project progress
Purchasing	540	45	49	22
Traffic	150	12	15	2
Expediting	160	13	17	2
Inspection	160	13	11	1
Liaison	200	17	30	5
	1210	100%		32%

The procurement function is 32% complete overall.

MEASUREMENT OF PROGRESS IN
PROJECT MANAGEMENT

Probably the most difficult aspect of project work to measure physical progress in, is project management. This may not be a problem on most small projects in operating plants in which project management is considered an overhead and is not charged to the project. However, on small projects that must "stand alone," the project management cost is apt to be significant enough to justify the effort to control it.

Our approach to defining a progress-measurement system for project management is the same as for all other types of work: to break the work down into measurable tasks or products. For project management, we can define the physical results of project-management efforts as including:

Progress reports
Cost estimates
Network plans
Schedules
Contracting plans
Purchasing plans
Quality control program

We can also define key milestones for each project-management function, which can also be linked to progress. We would then assign a value for percent complete to each milestone. Progress can be assumed to be linear between the milestones. For example:

1. For the overall project-management function:
   Authorization to proceed
   Award of design-engineering contract
   Completion of conceptual design
   Placement of major purchase orders (e.g., for long-lead equipment)
   Award of construction contract
   Completion of engineering
   Completion of construction
2. For the planning and cost-engineering function:
   Completion and approval of the budget plan
   Preparation of each cost and schedule forecast
3. For the contract and procurement function:
   Award of major contracts
   Placement of major purchase orders

Delivery of major equipment
Completion of major contracts
4. For the quality-control function:
Approval of Quality Assurance/Quality Control (QA/QC) plan
Inspection and approval of major fabrication items
Inspection and approval of major construction items
System commissioning and start-up

This milestone technique is discussed further in "Shortcut Techniques for Measuring Progress," later in this chapter.

PROVISION FOR CHANGES

A classic problem with progress-measurement systems is the problem of coping with changes. In most cases, if we measure progress against the original scope of work, we are apt to encounter periods in which progress is offset by increases in the scope of work. Larger projects have been known to experience negative progress in months in which a large number of changes have been approved. It has not been easy for those project engineers to explain to management how they were able to spend thousands of work-hours during the month and have -5% progress to show for it!

In the discussion of contingency presented in Chapter 6, we find that contingency is provided in an estimate and schedule for those variations that are likely to occur but that cannot be specifically identified at the time the estimate is prepared. It is, therefore, unrealistic to state that we are, for example, 50% complete with a project if that 50% refers to an original scope of work that is likely to be less than the actual scope. It is far more realistic to measure progress based on the current approved scope of work, i.e., the current model, which reflects the cost, time, and resources necessary to complete the approved scope.

SHORTCUT TECHNIQUES FOR MEASURING PROGRESS

Methods Based on Judgement

On many small projects—such as projects with short durations, or perhaps projects that are executed simultaneously with many others—a shortcut approach to progress measurement is sufficient. If used correctly and consistently, there is nothing wrong with shortcut techniques for progress measurement. Certainly, an approximate progress measurement is far better than no progress measurement at all.

In general, an approximate method for progress measurement must rely on judgement. The use of judgement, a time-honored method, has fallen into some disrepute as the marked tendency of such judgements to be optimistic has become more apparent. Judgements on progress often suffer from the fact that those whose performance is being measured can hardly be expected to be objective in making that measurement, and those who dispute the measurement have nothing more than their own opinion on which to base their arguments. However, progress measurements based on judgement can be a valid approximation if the basis for making that judgement is clearly defined and consistently applied. Once a method and guidelines are established for small projects they can, of course, be used on all projects and often end up saving time that would otherwise be spent in progress-related disputes. Some methods for using judgement in approximating progress are described below. In all cases, judgement should at least consider the earned value principle by always looking toward the physical manifestation of progress, as opposed to the time, effort or money spent.

Using the Network Activities for Progress Measurement

If a project is represented by a network, there will be some activities that are complete, some that are in progress, and some that have not yet begun. Therefore, as a percentage of the total job, the activities on which progress is to be measured often do not represent the major component. So using a network as the basis for our judgements on progress is somewhat more valid than just looking at the project as a whole. The accuracy and consistency of the progress measurement will be enhanced further if we use milestones as described below.

Milestones for Progress Measurement

If our projects tend to follow a similar pattern, it is possible to establish a set of agreed-upon guidelines tied to milestones. For example, if our projects all involve small buildings, we might set up a system as follows:

Activity	Percent of project complete when activity is complete
Survey	5
Prepare plans	15

Activity	Percent of project complete when activity is complete
Obtain financing	20
Obtain permits	25
Grading and excavation	30
Pour foundation	35
Frame building	45
Begin plumbing and wiring	50
Roof building	60
Install siding, windows, and exterior finish	70
Install interior carpentry	80
Finish plumbing and wiring	85
Finish interior	95
Final grade and landscaping	100

A set of guidelines based on these milestones should document what is meant by each activity's completion. For example, it should be clear exactly what work has been done when the framing building step is complete, and what work is included in each succeeding activity. The allocation of percentages can then be based on our judgement of the allocation of work-hours or costs. For progress measurements on a given day, judgement would be used to interpolate between milestones or handle situations in which several incomplete activities are in progress at once.

"Remaining Duration" to Measure Progress

When progress is measured by judgement it is best if that judgement's format relates well to the way that people think. For example, when asked, "What is your % complete?", many laymen will give an optimistic answer, not only because of their naturally optimistic bias but also because of their tendency to minimize in their minds the work remaining. However, if a field superintendent or foreman is asked, "How long will it take you to finish this work (given a certain number of people, materials and equipment)?", he or she can usually give a pretty accurate answer. The remaining duration of the activities in progress can then be used to update the project network and to forecast the project-completion date. The % complete of each activity is calculated as the ratio of time spent to date over total duration. Overall % complete can then be calculated in the usual way by weighting the % complete of each activity by its planned work-hours.

Use of Physical Quantities to Measure Progress

Some companies find it useful to use overall physical quantities to measure progress. For example, on a job with a lot of welding, we might use cubic in. of weld installed as a progress indicator. Knowing the total cubic in. of weld metal to be laid down, and the amount done to date, we have a rough indication of progress. Similar procedures could be followed with other physical quantities, such as in.-ft. of piping installed, tons of structural steel erected, linear ft. of cable installed, etc.

Unit-price contracts, by nature, provide a good progress measurement method, as they require physical quantities of work to be measured accurately if the contractor is to be paid. Therefore, in this case, the progress-measurement system should be compatible with the schedule of unit prices, such that the quantity measured for contract administration gives us an automatic input for progress measurement.

CHAPTER SUMMARY

Effective project control requires some measurement of physical progress. An understanding of the basic principles of the earned-value system is essential even if progress measurements based on judgement are to be used. For small projects, approximate methods can be used effectively provided the guidelines are clearly defined and consistently applied. Planning networks provide a good basis for a number of shortcut progress-measurement techniques. The project model provides the basis for weighting the progress on various activities in order to calculate overall progress.

13

Forecasting and Control of Cost, Time, Resources, and Quality

INTRODUCTION TO FORECASTING AND CONTROL

"There is no such thing as Cost- or Schedule-Control."

This statement might seem quite controversial to those who make their living doing just that—controlling the cost and schedule of projects! But it is, nonetheless, true. We cannot control what it costs to do something, nor how long it takes. *What we can control are those things that determine what the cost and time will be.*

By the time we know what something costs, it is much too late to control the cost. And by the time we know how long it took to do it, it is too late to control time. Therefore, effective control of these variables requires forecasting, so that control actions can be taken in sufficient time to avoid cost overruns and schedule delays.

This basic principle underlies all that there is to say about project control. As we saw in Chapter 11, it is a matter of knowing three data points (see Figure 13.1):

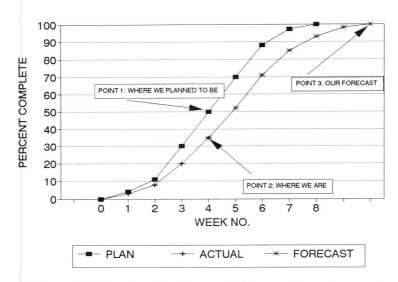

Figure 13.1   A project is controlled by tracking and comparing differences between three points: the plan, actual performance, and a forecast at completion.

Where we planned to be (e.g., 50% complete)
Where we actually are now (e.g., 35% complete, or one week behind
      schedule)
Where we will end up if present trends continue (e.g., two weeks late)

When we know where we are versus where we planned to be, we know whether or not we have a problem. (For example, we might know that we are one week behind schedule.) But before we can make an intelligent forecast, we need to know *why* we are behind schedule—and what should be done about it. To determine why a cost or time variation has occurred, we must be able to analyze those factors that determine the cost and time such as:

Productivity
Timely receipt of information and materials
Interferences
Number of resources available

The tracking curves and checklists described in Chapter 11 are the tools for performing the analysis that answers the "why" question, and the earned value calculations described in Chapter 12 provide the required objectivity.

CREDIBLE FORECASTING—A KEY TO CONTROL

Once a variation from the plan has been identified, along with its cause, it is possible to prepare a forecast of the time, resources, and cost to complete the project. This forecast will be used to justify the corrective actions that need to be taken in order to correct the variation and recover the cost or schedule.

Therefore, we can see that the purpose of a cost and schedule forecast is not always to predict the project's completion date and final cost. Rather it is to *predict the completion date and final cost that will occur if nothing is done to correct the problems identified to date.*

Why make such a pessimistic forecast? After all, don't we expect that the recommended corrective actions will be taken and will be effective? Well, in the world of multiple small projects, this may not always be so easy.

Let's look at an example: Suppose the Information Systems Department has 50 small projects in progress. We identify that Project 32 is behind schedule by two weeks because increases to its scope of work have caused it to require two more programming staff members if it is to be completed on time. These additional two people, of course, will have to be taken from their work on projects 12 and 26, thereby delaying these other projects. Or, contract programmers may have to be brought in, requiring extra expense and effort.

Chances are the I.S. Department manager will not look favorably on this prospect. She may express some skepticism that the resources already assigned to our Project 32 are inadequate to finish on time. She may express frustration that solving the problem on one project only causes more problems for others.

As the leader for Project 32, we can use a credible forecasting technique to provide the manager with a clear set of alternatives: Provide two more programmers for Project 32 or accept a likely delay in completion of two weeks. Note that we do not intend this as a threat, but simply a realistic choice of alternatives. In the multiple project environment, each project's leader will naturally assume that his projects are the most important. Yet, there is an issue of priorities that a manager can address, and credible forecasting makes the choices clear. We can say that the forecast final variation indicates the *value* of taking the recommended action.

INTERACTION OF COST, SCHEDULE, AND RESOURCES

In "Integration of Cost, Time, and Resources," in Chapter 2, the concept of integration was introduced, and the inherent dependencies of cost, time, and resources explained. Nowhere is this dependency more noticeable than in the area of project control. We can easily observe that if we seek to tighten our schedule, the costs will probably increase, as will the amount of manpower required, while quality is apt to suffer. If we fail to provide sufficient manpower resources, both schedule and cost are likely to suffer.

Effective project control, therefore, requires an integrated approach. Fortunately, our project model provides us with the ideal basis for integrated project control. In Chapter 2 we saw that our project model tells us, for each activity:

What scope of work is performed
What material, manpower, and equipment resources are required
How much time the work is expected to take
What it will cost (to perform that scope of work, with those resources, in that period of time)

As the project progresses, the actual scope, the actual duration, the actual dates, the actual resources applied, and the actual costs will all be somewhat different from the plan. The reasons for these differences are the project trends and developments that we seek to control.

From the previous two chapters, we can define the essential elements of an effective project-control system:

A means for timely management reporting
A progress-measurement system
A basis against which to compare actual costs, times, resources, and scope of work

For multiple small projects, these essential elements may, if necessary, be provided in a rough, approximate way. The important thing is that each be provided. Often methods can be set up that are applied consistently on each small project, so that the group of multiple projects remains under control. See Chapters 14 and 15 for software features that aid in "standardization and customization."

BASIC PRINCIPLES OF FORECASTING AND CONTROL

A definition of the proper basis of a cost or schedule forecast was provided in "Using the Tracking Curve for Forecasting and Control" in Chapter 11, and is worth repeating here:

A forecast predicts the final outcome of the project if the trends identified to date continue to the project's completion. Our cost/schedule forecast is the way in which we focus the "spotlight" on the areas requiring management attention.

To make a forecast and identify problem areas, we need to ask some important questions. How well we answer these questions will determine how well we control the project. The questions that need to be asked with regard to schedule; design and construction quality; labor, material, and equipment resources; and costs are:

Where did we plan to be at this point?
Where are we now?
If there is a difference between our planned and actual situation, what is the reason?
If things continue as they are, what will be the result?
Has corrective action, applied to problems previously identified, been effective?
What are current problems and what can be done to correct them?

We will see specific examples of this measuring, comparing, analyzing, and forecasting process later in this chapter.

SCHEDULE FORECASTING AND CONTROL

Assessing Current Schedule Status

One way to assure that we stay "on top" of the project is to regularly review its status. It is helpful to conduct these reviews by using a checklist. A checklist of the project-control variables that affect the schedule is as follows:

Physical progress to date
Productivity (the relationship between physical progress and actual workhours)

Maintenance of schedule milestones
Availability of materials
Availability of labor
Provision of overhead services (supervision, transport, etc.)
Progress and productivity trends to date

In order to accomplish schedule control we must evaluate how the project stands with respect to each of these variables in comparison to the original budget plan. The more complete our plan is, the better we will be able to identify specific areas of variation and specific reasons why the variation has occurred. For example, if we have an integrated plan that shows the schedule for the application of specialized manpower, we can quickly ascertain whether any delay in progress was due to inadequate manpower by comparing planned versus actual utilization (see Figure 13.2). Such a comparison also indicates which time periods were involved.

As a minimum checklist for small projects, we should ask ourselves:

Is progress as planned? If not, why not?
Are the work-hours expended as planned? If not, why not?

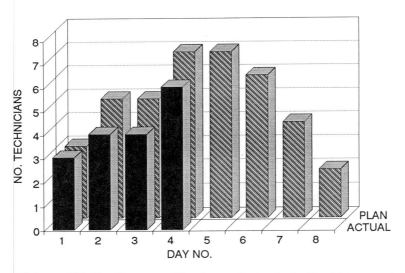

Figure 13.2   Resource utilization can be tracked with a histogram showing planned versus actual numbers of resources provided.

Are materials being shipped and received on schedule?

Are major milestones still on schedule?

Have any major problems been identified? If so, are they being handled properly?

As before, our answers to these questions point the way to the actions required for effective control.

Forecasting Schedule for Completion

Once we have completed our review of current schedule status, we must prepare an updated forecast of the schedule for the work remaining. To do so, we must reexamine all the assumptions behind our original schedule, and redefine the basis for the work still left to do. It is important to remember that the schedule forecast must be consistent with the current schedule for the application of manpower, equipment, and materials resources.

There are various methods for schedule forecasting. The appropriate method for a small project depends on the level of detail of the plan, the importance of the project, and the time available for forecasting and control. Some methods are discussed below.

Updated Network and Resource Analysis

If the original plan and the project-control methods are computerized, it is a relatively simple matter to run the scheduling program for the remaining scope of work. The basic steps to prepare the updated schedule are as follows:

1. *Check the network logic.* Mandatory dependencies ("hard logic") will probably still be valid, but arbitrary dependencies ("soft logic") may well be subject to change. In the case in which the schedule has slipped but the completion date must be maintained, it is often useful to reexamine the dependencies. There may well be activities that can be done in parallel, although it may be more inefficient and costly to do so.

2. *Review past and future milestones.* If past milestone dates have slipped, future milestone dates may also be likely to slip unless the cause for the previous slippage has been identified and corrected.

3. *Check the delivery dates* for materials required for the work left to go. If delivery dates have slipped, check network float to see if slippage can be accommodated without changing the schedule.

4. *Check the availability of resources* as required for the work left to go. If manpower shortfalls are likely, revise activity durations accord-

ingly and/or update the resource analysis to reallocate manpower between activities.

5. *Input actual start and completion dates* for activities that have been completed.

6. *Input actual progress* and remaining durations for activities that have begun.

7. *Check the Critical Path.* Remember that the length of the critical path sets the length of the project. Always check progress to date, resource requirements and availability, and potential changes along the critical path. Be careful to realistically forecast the time to complete critical activities.

Use of Average Rate of Progress

On small projects for which an in-depth schedule analysis may not be justified, we can simply calculate the average rate at which we have made progress to date (e.g., 10% per week) and assume a rate of progress to the end of the job. Such forecasts should use the typical S-shaped progress curve, and provide for a slower rate of progress towards the end of the work.

Use of Milestones

On small projects without a proper plan and schedule, milestone dates can be used to indicate the likelihood that the planned completion date will be achieved. As mentioned above, if the planned milestones to date have slipped, chances are the final date will slip also unless the causes are identified and corrected.

Use of Tracking Curves

The use of tracking curves to define current status and forecast the work left to go is discussed in Chapter 11.

A short-cut method for forecasting time, resources and costs to complete the project is with the use of simple formulas. The earned value method described in Chapter 12 provides us with useful data that can be used in formulas that often help convince people that a forecast is reasonable.

From our current project model we know: The Total Work Hours (TWH) to complete the current scope of work of the entire project, and the Planned Work Hours (PWH) that should have been earned to date.

From our progress report we know: The Earned Work Hours (EWH) (physical progress × planned workhours), the Actual Work Hours (AWH) spent, and the physical progress "% Complete"

$$\% \text{ Complete} = \frac{\text{EWH}}{\text{TWH}}$$

Productivity, or "Cost Performance Index" (CPI), is calculated as:

$$\text{CPI} = \frac{\text{EWH}}{\text{AWH}}$$

and, the Schedule Performance Index (SPI) is calculated as:

$$\text{SPI} = \frac{\text{EWH}}{\text{PWH}}$$

We now forecast the total workhours spent at completion of the project as follows:

Forecast Total Workhours Spent at Completion ("FWH")

$$\text{FWH} = \frac{\text{AWH}}{\% \text{ Complete}}$$

or:

$$\text{FWH} = \frac{\text{TWH}}{\text{CPI}}$$

We can also forecast the workhours required to complete the project:

WORKHOURS TO COMPLETE $= \text{FWH} - \text{AWH}$

or:

$$= \frac{\text{TWH} - \text{EWH}}{\text{CPI}}$$

If we know the number of resources available to the project, we can also forecast the minimum time to complete the project, assuming all resources can be used productively each day:

$$\frac{\text{MIN. TIME}}{\text{TO COMPLETE}} = \frac{\text{WORKHOURS TO COMPLETE}}{\# \text{ RESOURCES AVAIL.} \times \text{HRS/DAY}}$$

FORECASTING AND CONTROL OF RESOURCES

Importance of Resource Management to Small Projects

As discussed in Chapter 4, small multiple projects are particularly sensitive to the cost and schedule effects of resource shortfalls because of the competition for resources, as well as the limited number of alternative ways to make progress. If, on a large project, labor, materials or equipment are not provided as planned, it is often possible to make progress on other parts of the project where resources are available. This degree of freedom is not available on a small project, and often the lack of critical resources can stop progress altogether. Planning, tracking, forecasting, and controlling the utilization of resources is therefore of particular importance on the small project. As in other aspects of project control, the project model provides the basis for control.

Small projects may obtain design services, materials, labor, and equipment either in-house or from outside contractors and suppliers. In some cases, a single project will have resources supplied from both inside and outside sources. Although it generally is someone else's job to do the purchasing or contracting, it is the project leader's responsibility to see to it that these resources are indeed provided. This supervisory task is often discharged more effectively when a clear and specific schedule is agreed upon in advance, and then used as a basis for tracking, forecasting, and control. The resource schedule is described in Chapter 4. As a minimum, the project leader should check that:

The quantities and types of materials, labor, and equipment that are currently required are on the job

The quantities and types of materials, labor, and equipment that are required for the upcoming work will be provided

Any problems relating to materials, labor, and equipment have been identified and corrective action has been taken

Forecasting and Control of Labor

Labor falls, in general, into two categories:

Direct labor (labor that results in physical progress)
Indirect labor (project specific but not direct)
Overhead labor (not project specific)

As illustrated in Figure 13.2, we can use the resource schedule provided in the current project model as a basis for tracking labor resources. At any point in the project, we can update the resource-analysis process by which we optimized the original time and resource plan to reflect the work-hours spent to date, the physical progress to date, the current schedule for the work remaining, and the anticipated availability of labor for the remaining work (see Figure 13.3). It should be noted that changes to the schedule for the work left to go will have an effect on the requirement for labor resources.

When updating the schedule for the provision of labor we need to consider the following factors:

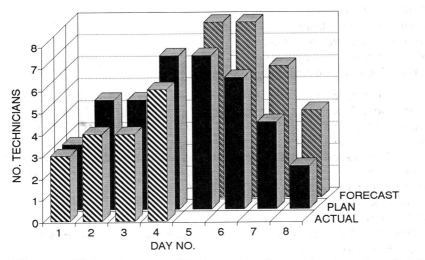

Figure 13.3 Resource utilization can be forecast based on the schedule for work remaining, and shown in a histogram.

Physical progress to date, which indicates how much work remains

Labor productivity to date, which indicates the manpower required to perform the remaining work

Contractor performance in providing labor, which indicates whether we need to seek alternate sources of labor

Schedule of work to go, which indicates when the labor will be required

Scope of work to go, including approved and anticipated changes

Control of labor can be thought of as tracking and forecasting the quantity and quality (i.e., productivity) of the manpower applied to the job. In the multiple project environment this is especially important as each project is counting on others to release needed resources on time.

Forecasting and Control of Equipment

On a small project it is generally not necessary to schedule and control all types of equipment. In general, such equipment as trucks, welding machines, small tools, etc., are readily available. (Projects in remote locations are, of course, exceptions). The type of equipment that requires some planning and control has one or more of the following characteristics:

Is not readily available
Is critical to the schedule
Requires some time to mobilize
Interferes with other operations
Is shared among multiple projects

For example, our small project may involve lifting and placing a compressor. This activity, on the critical path, requires a large crane. The crane has a great cost and schedule impact on the project since, if the crane is not available when needed, the entire project will be delayed, and, if the crane stands idle even briefly, the costs will overrun. The crane's operation must also be carefully planned to avoid interference with existing facilities and operations. If the schedule should change (due, for example, to a delay in shipment of the compressor), changes to the arrangements for the crane must also be made quickly if charges for idle time are to be avoided. This crane is the type of equipment that requires scheduling and control.

The techniques for forecasting and control of equipment resources are much the same as those described above for labor. As the schedule is updated, the requirements for and availability of equipment should be reviewed, and the equipment schedule revised accordingly.

Control of Materials

The various operations involving the provision of materials to the job are generally handled by the in-house procurement function, or by the contractor. These operations include purchasing, expediting, inspection, traffic, and warehousing. The project leader's role is to coordinate and supervise these activities. Unlike labor and equipment resources, materials are an "all or nothing" proposition: they are either there or they are not. Until they are actually on the site, the project leader must rely on information from various sources to figure out what is going on, and what to do so he can assure that the material will be there when required. Often that information is a "pink copy" carbon of a purchase order, memo, or receipt notice: hardly the kind of timely information that is required. What is required is a simple but effective method for capturing and presenting information about the status of materials so the project leader can take action if required.

Planning for Materials

Materials can be divided into three categories according to their importance to the schedule:

Long-lead materials
Critical materials (those materials are on the critical path and may or may
    not also be long-lead items)
Materials with float

    The first step is to identify the materials in each category, which can
be easily done using the schedule-analysis principles described in Chapters 3 and 4. The next step is to recognize that material procurement has
a schedule of its own—a subnetwork to the project schedule—and that
all materials follow this plan, as shown in Figure 13.4. From Figure 13.4
we also note that material purchased for the project requires more time
than material that can be obtained from the company warehouse. The
project leader's job then becomes one of tracking where each material
item is on its procurement schedule, starting with the critical items, then
the noncritical long-lead items, then the items with float. It might also be
helpful to prepare a separate schedule showing the key milestone dates
for placing the purchase orders and for delivery of materials, and use
this schedule for discussions with the procurement function, as well as to
track progress in procurement.

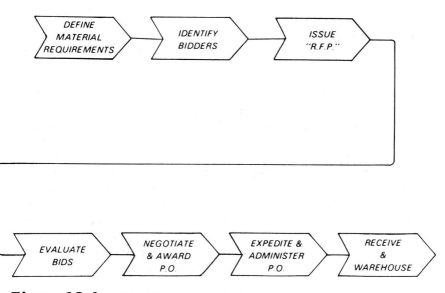

Figure 13.4   Material procurement plan.

If project control is computerized, it is possible to set up a database, possibly tied to the network, which presents an up-to-date summary of material status. The material data can be linked to the network by specifying the required material as a resource requirement for the appropriate activity, and the computer system can then link the network activity to the procurement information in the database. The key to the success of such a system is to capture data from existing documents, and to avoid creating new procedures.

Controlling Materials

For the small project, for which we are not likely to have a sophisticated material-control system, we can use a simple checklist to help control the materials function:

Have design specifications and purchasing requisitions been completed on schedule?
Have purchase orders been placed on schedule?
Has everything that has been requisitioned to date been ordered?
Have the correct items and quantities been ordered?
Are the anticipated delivery times in accordance with the schedule?
Are immediately-required materials available?
Does the schedule for purchase and delivery of materials reflect the latest update to the project schedule?
Do the current material requisitions reflect the latest design changes?
Have problems related to materials been identified and corrective action taken?

As the materials status is periodically updated, the overall project schedule must, of course, be revised accordingly.

One of the interesting uses of float in a schedule is in assessing the impact of anticipated delays in the delivery of materials. On a project without a network plan and schedule, delays in materials are often interpreted as a problem requiring immediate corrective action. Time and money are often spent devising ways to improve the delivery times. In some cases, however, a network-based schedule will show that the material delivery slippage can be accommodated within the existing float, and that corrective action is not necessary. Therefore, network time analysis helps not only by focusing attention on those items that are critical, but by avoiding panic situations over items which are not.

Allocating Fixed Resources to Multiple Projects

In the small-project environment, in which we often have many projects underway simultaneously, the main problem involving resources is often that of allocation between projects. For example, the company may have a Drawing Office that does the design and drafting work for each of the small projects, and the problem is to allocate the limited resources of that office to the various projects. Or, the project-engineering department itself may consist of a number of project engineers, each of whom handles several projects, and the problem for the department manager becomes, "To whom should I assign this new project?" Often the company has a pool of labor forces and construction equipment, and these resources have to be allocated among the various projects. This problem is discussed in Chapters 4, 7, 8, and 15.

Some companies find it helpful to use a priority system for small projects. This often works well, and is useful to implement priority-driven resource scheduling (see Chapters 4, 7, and 15), but the problem with such systems is often that lower-priority jobs are never done until they slip so badly that their priority is upgraded. This undermines the priority system and tends to make it based more on office politics than on the actual priority of the projects.

The concept of the network hierarchy was introduced in Chapters 2 and 3. This technique, in which we summarize several activities and create a higher-level network, allows us to create a "project" that is, in effect, the work required to do all the projects currently defined. Since all the projects shown as activities cannot be executed simultaneously, we can schedule them according to the availability of resources. Using resource-limited scheduling, we can prepare a resource allocation schedule. As new projects are added, current projects slip, and existing projects are completed, the Level 1 schedule is changed accordingly. This procedure requires a computer system, but, assuming one is available, provides a most effective way to handle allocation of fixed resources to multiple projects (see Chapters 4, 7 and 15).

COST FORECASTING AND CONTROL

What is "cost control"? To some it is the recording and analysis of cost data such as timesheets, purchase orders, and invoices. Such data is historical, and, as such, it makes an essential contribution to cost control. However, no one can control anything solely by concentrating on what has already happened. We will refer to cost control in the context of the things we do to affect the current and future activities of the project, and, hence, the final outcome.

We cannot control costs, but we can control the things that happen on a project to determine its cost. We can assure that the cost impact of the various decisions that are made is recognized as part of the decision-making process. We can assure that unnecessary costs are not incurred due to poor productivity, schedule slippage, inefficient allocation of labor resources, unnecessary design or field changes, and so forth. We can, finally, assure that we have a well thought out and documented planning and cost basis for use in tracking and controlling the project, as well as the appropriate methods, systems, and procedures. In fact, all the material presented so far has to do with cost control.

Cost control, then, consists primarily of assuring that the project is managed with full recognition of the cost impact of everything that is done, and everything that happens. In this section we will review some basic principles and aspects of cost control.

Analyzing Variations to the Estimate Basis

In "Establishing the Estimate Basis," in Chapter 5 the importance of the estimate basis was discussed. The estimate basis can be thought of as a "tripod," the three "legs" of which are the design basis, the planning basis, and the cost basis (see Figure 5.2). The basis, therefore, describes what is to be built, how it is to be built, and the pricing levels that are anticipated. All variations between the actual costs and the estimate can be explained in terms of variations to this estimate basis. In "Reconciling Forecasts and Estimates" (Chapter 5), the use of reconciliations was discussed as a control device in which the variations to the estimate basis were identified and costs attached to them to explain the differences between actual, estimated, and forecast costs.

One aspect of cost control, then, consists of periodic reviews of the project as it progresses, and the comparison of these reviews with the estimate basis. We might use a checklist for such reviews, as shown below. As we review each item we should ask: "How does the way the project is actually progressing differ from the assumptions made in the estimate basis? If there is a difference, what is the cost effect likely to be?"

The design basis
  Specification of overall performance (e.g., capacity, performance characteristics)
  Overall scope of work
  Type and extent of design changes
  Type and extent of field and startup changes
  How late in the project design changes are made
  Work-hours spent in design work
  Extraordinary technical problems identified
  Timeliness and effectiveness of design reviews
The planning basis
  Schedule duration
  Milestone dates
  Contracting plan
  Purchasing plan
  Activity constraints
  Use of shift work and overtime
  Manpower density
  Progress relative to reference schedule
  Labor productivity
  Engineering productivity
  Availability of materials
  Availability of labor
  Availability of construction equipment
  Availability of drawings
The cost basis
  Prevailing escalation rates
  Average hourly costs for engineering
  Average hourly costs for labor
  Average unit costs for materials (e.g., dollars per ton of piping)
  Quantities of materials
  Number and cost of changes
  Contingency required
  Indirect costs as a percentage of direct costs
  Bulk materials as a percentage of equipment costs
  Cost to achieve 1% physical progress

Tracking curves, discussed in Chapter 11, are a useful way to monitor trends and identify variances. Some additional sample tracking curves are shown in Figure 13.5. These computer-generated curves show information presented in cumulative and histogram formats. Extrapolations from the tracking curves can be used for cost forecasting.

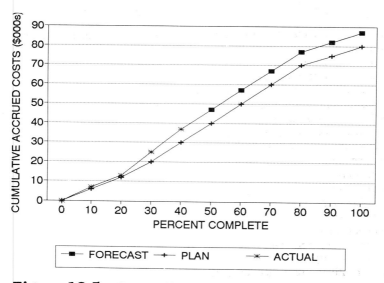

Figure 13.5   Cost tracking and forecasting. Cumulative cost vs. percent complete.

Cost Forecasting

The same rules of forecasting that were described in "Basic Principles of Forecasting and Control" earlier in this chapter also apply to costs. A cost forecast should reflect the expected final cost if things continue as they have been, and nothing is done to correct negative trends. If we are using an integrated cost-schedule-resources project model, the "current model" represents design changes approved to date, the "forecast model" represents what is expected to happen, while the "static model" represents the budget plan and the estimate basis.

In cost forecasting, it is important to distinguish between factors affecting the project that are controllable and those that cannot be controlled or corrected. These might be called "external" or "internal" factors, depending on whether they impact the project from an external cause (and therefore are not subject to the control of the project leader), or whether they are internal to the project. For example, escalation rates cannot be controlled, although it may be possible to take some steps to minimize their impact. Design changes, on the other hand, can be controlled. Labor productivity may or may not be controllable, depending on whether it is determined primarily by market conditions (and is therefore external), multiple-project resource allocation problems (internal), site conditions (internal) or contractor performance (internal). The net cost impact of the external, uncontrollable factors establishes a minimum level for the cost forecast.

In general, cost forecasts are made by using the project model to calculate the value of work done, and forecast the cost of the work remaining. A forecast can also be made by updating the estimate basis for productivity, wage rates, etc., and recalculating the entire estimate. A key point about cost forecasts is that they are a matter of judgement, and the forecaster should feel free to make the forecast according to his best analysis and judgement.

Control of Changes

One of the most frequent causes of cost overruns is changes. This is most unfortunate because changes are generally an internal, controllable factor. Therefore, one of the most important things that can be done to effect cost control is to establish proper control over changes.

Definition of a "Change"

What is a "change?" Many spirited discussions have taken place over just that question, particularly when contract terms cause large sums of money to ride on the answer. We can define a change as follows:

A "change" is a specific work assignment which would not ordinarily be assumed to be required to complete the original scope of work. A change may also be an instruction to perform a specific work assignment in a different way from the way which was previously defined.

Changes are normally made to improve the performance, operability, maintainability, safety, or cost of the facilities.

Note that this definition excludes "design development" work from the category of changes. As design work progresses and the design is defined in increasing detail, alternative approaches to various design problems are studied, and a number of revisions to the work already done normally occur. This is a normal process, and these minor revisions are not generally considered changes, even if they are initiated by the client in the course of normal reviews. If, however, an instruction is given to do the work in a different way than might normally be assumed, or to do an item of work not normally required to complete the original scope of work, then that instruction requires a change to be approved.

Now that we have defined the basic concept of a change, we can define the different types of change (see also "Definition of Project-Control Terminology," in Chapter 11).

Design change: A change made during the design phase of the project which modifies work already done or adds work not normally assumed to be required to complete the original scope of work. A design change is initiated by the design function.

Scope change: An increase or decrease to the original scope of work. This generally means a change to the overall specification or objectives of the project (e.g., capacity, facilities installed, acres of land, etc.) or to a contractually-defined scope.

Field change: A change initiated in the field to facilitate construction

Startup change: A change made to facilitate or simplify the startup of the facilities. Also, a change made during the startup phase of the project.

Control of Changes

Most engineers have a tendency to change things; always, of course, to make them better. Changes are inevitable on a project and they are often initiated faster than we can keep track of them. Unfortunately, once changes get out of control, our cost-control work becomes ineffective because we can no longer analyze variances since we cannot distinguish those variances caused by changes from those caused by other factors.

Changes are often difficult to estimate as their cost impact tends to have a "ripple effect." A change that seems to affect one department may have a larger effect than anticipated. For example, suppose the horsepower rating of a pump is increased. In addition to the pump costing more, its driver will cost more too. The pump and driver will weigh more, possibly resulting in increased weight and cost of structural steel and foundation. The piping from the higher-power pump will likely be of a higher flange rating, resulting again in more weight and cost for the piping and its supporting steel and foundations. Similarly, the electrical cables to the pump will be larger and heavier, as will the switchgear and other electrical equipment. Additional work-hours will be required in the field to install this heavier equipment and materials, and additional work-hours required to revise the design. As a result, the engineer who suggested a change to the higher-hp pump, which might cost $5000 more, is often shocked to find that the net cost impact is $50,000 when all the costs are accounted for.

The timing of a change also has an effect on its cost. As seen in Figure 13.6, a change made early in the design phase, when everything is on paper and is still preliminary, will have the minimum cost. As the design progresses, more drawings in more departments must be changed to accommodate the same design change. Finally, after construction has begun, the change might involve removal or modification of construction work already done, and that involves the highest cost.

The essential elements of a change control system are:

The ability to identify changes as they occur
The ability to prepare quick estimates of the cost impact of each change
The requirement that a manager or project engineer with responsibility
    for the budget approve changes only when the cost impact is known.

To create such a system, we need formal procedures and a quick estimating method. It is also helpful to define the various stages of a change, so that they can be tracked through each stage.

Potential change: A change that is still in the idea stage, but which is likely
    to be initiated.

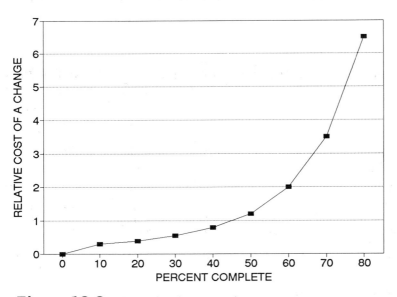

Figure 13.6   Cost of a change vs. time.

Pending change: A change for which approval is sought but has not yet been obtained. A change is pending during the time that the design details are worked out and the estimate prepared.

Approved change: A change for which the design and estimate have been approved. Approved changes are usually included in the "current control estimate," and are offset by reductions in contingency.

Cost Reporting

In general, cost reports should have the basic elements provided in the example shown in Figure 13.7 and described below:

Cost category: This groups costs into categories for control.
Budget estimate: The budget for the project (the "static" model).
Approved changes: Shows the total approved changes to date.
Current control estimate: This is the budget estimate plus approved changes, and is the "current model."
Current cost forecast: This is the value of work done plus cost of work to go, and is the "forecast model."
Variance: This is the current forecast—current control estimate.

Note that the current cost forecast includes provision for pending changes as well as for contingency to cover the uncertainties surrounding the re-

COST CATEGORY	BUDGET	APPROVED CHANGES	CURRENT ESTIMATE	FORECAST FINAL COST	VARIANCE $	%
			(thousands)			
DIRECT COSTS						
DESIGN	100	5	105	110	5	4.8%
MATERIALS	175	7	182	190	8	4.4%
LABOR	200	17	217	245	28	12.9%
TOTAL DIRECTS	475	29	504	545	41	8.1%
INDIRECT COSTS						
TEMP.FACILITIES	50	2	52	55	3	5.8%
SUPERVISION	150	3	153	145	-8	-5.2%
HOME OFFICE	75	4	79	85	6	7.6%
EQUIPMENT	90	6	96	99	3	3.1%
SUBTOTAL INDIRECTS	365	15	380	384	4	1.1%
TOTAL BASE ESTIMATE	840	44	884	929	45	5.1%
PENDING CHANGES						
CONTINGENCY	130	-35	95	90	-5	-5.3%
TOTAL	970	9	979	1019	40	4.1%

Figure 13.7   Typical cost report format.

maining work. Attached to the cost report might be a list of approved, pending, and potential changes, tracking curves as appropriate, a written analysis of cost trends, and a reconciliation to the previous forecast as well as to the budget.

QUALITY CONTROL

Although quality control is a highly developed subject, the details of which are outside the scope of this book, there are a number of basic principles that are relevant to the project leader. For the purposes of this book, we will use the following definitions.

Quality: Conformance to requirements. The better something fits the purpose for which it was intended, the higher quality it is. In most cases, the intended purpose of an engineered facility is to produce a profit for the company. Therefore, the item of highest quality will not always be that which is most expensive as it will be that which represents the optimum combination of price and performance (see Chapter 10).

Quality assurance: This is the standard we use to specify quality and the program of design reviews, procedures, and inspections we intend to use to assure adherence to those standards.

Quality control: This is the activity which we conduct to carry out the quality assurance plan.

Therefore, in the small-project environment, we set up a quality-assurance program for all small projects, and conduct quality control on each one. The basic activities involved in quality control on a small project are:

Definition of requirements in which the project leader works together with the "customer" to assure that the most cost-effective requirements are defined (see Chapter 10).

Design reviews in which the operations and maintenance personnel who will operate the facilities make comments and suggest changes to the design. The purpose of such reviews is to incorporate the design input of the user at an early stage in the design process to minimize the cost of such changes, as well as to assure adherence to standards and specifications.

Inspection and testing in which shop- and field-fabricated equipment and facilities are tested and checked.

Certification in which approval is obtained from a certifying authority (e.g., nuclear and marine projects).

Change control as described above.

We can use our project model for quality control by using it to identify, plan, and schedule the specific quality-control activities required. We can then monitor actual performance to assure that the required inspection, testing, and other quality-control activities are performed. If our system is computerized, we can set up a simple database to keep track of the relevant documents to assure that the proper certificates can be produced at any time.

Given the broad definition of quality we are using, we can see that the capital and operating cost of the project have a great deal to do with its profitability and hence with its quality. Of the three legs of the estimate basis (design, planning, and cost), design has the greatest impact on project cost, and it is completely controllable internally. Therefore, good control of changes and close monitoring of the cost implications of design trends and decisions is an effective form of quality control.

CHAPTER SUMMARY

In this chapter we discussed the various aspects of project control. We saw how the integrated-project model provides an excellent basis for control and how important resources—labor, materials, and equipment—can be monitored and controlled. Much of the effectiveness of the project-control function relies on the planning, scheduling, and resourcing work done prior to the start of the project. This work makes it relatively simple to track performance, identify problem areas, and make forecasts. In the multiple small project environment, PC software can be used to set up standardized, automated procedures for tracking and reporting the status and forecast for each project.

V
COMPUTER APPLICATIONS FOR MULTIPLE PROJECT MANAGEMENT

14

Information Systems for Managing Multiple Small Projects

WHY CONSIDER COMPUTERIZATION?

Computers have become such an accepted part of our everyday lives that it becomes easy to lose perspective on the fundamental issues of why they are appropriate in a given situation, what they should and shouldn't be expected to do, and how to achieve the computer capability we need. These questions have become markedly more complex in recent years as the computer industry has offered many new possibilities while raising new questions.

Fortunately, in spite of the many new developments in computing, the fundamentals of system design, development, and implementation have not really changed. Equipped with these fundamentals, we can use the latest computer technology in an effective way, regardless of how rapidly or dramatically the technology changes.

This chapter covers the fundamental concepts of system design, development, and implementation as they apply to multiple project management. Like most other chapters in this book, this one is not intended to make the reader an expert. Rather, it is intended to give a project leader all he needs to know to be able to assure that effective computer-assisted project-management techniques are available for each small project. We will, for that reason, deliberately retain the perspective of the computer system *user* throughout.

Our first consideration as users should therefore be: "why consider computerization in the first place?" If we look at the most effective computer systems, and consider inherent capabilities offered by computers, we find some or all of the following general conditions for effective information systems:

A considerable amount of data that needs to be handled
A requirement to file data and also present selected data when needed
A short time available to process data
A relative simplicity and repetitiveness of operations for handling data
A large number of people in the organization involved with the data going into the system, the information presented by the system, or both
The consideration that the availability of timely and accurate information is essential to the achievement of business goals.

It is certainly evident from the above that the multiple small-project management situation fits all of the criteria for effective computerization. We have a lot of data to handle in a short period of time, a lot of people involved, a need to be able to store data for later use, a lot of data needing manipulation using simple and repetitive calculations, and, most important, a need for timely and accurate information. As we have seen in the preceeding chapters, computer power can be a time saver in:

Network analysis and scheduling
Resource analysis
Cost estimating and expenditure forecasting
Management reporting
Decision making

So computer-assisted techniques deserve close consideration in the multiple small-project environment. Let's begin our discussion with a definition of terms.

DEFINITION OF TERMS

As in other subjects covered by this book, it is worthwhile to define the terms we will be using.

Project-management system: A computer-system for processing project data in order to provide information to facilitate project-management actions.

Project-management method: A series of specific actions to accomplish a project-management task. The method may include use of a project-management system. The method is documented in a procedure that is followed uniformly on each small project.

Data: Numbers and words that represent some aspect of the project. "Raw" data is in the form in which it was captured, "processed" data has been subjected to some manipulation.

Information: The presentation of processed data in a way that conveys specific knowledge that will enable the person receiving it to perform project-management functions. Information can be presented in printed reports, with graphics, or on a screen display.

Input: The entry of raw data into the system. Input may be by means of a keyboard, disk, or electronically through a communication network.

Output: The presentation of information prepared by the system. Output can be in the form of printed reports, management graphics, a screen display, an electronic message to another computer, and/or a file on a disk or tape.

Random-Access Memory (RAM): The part of a computer in which data is processed. The amount of RAM indicates the size of the program and the amount of data that can be handled in memory. The contents of RAM are temporary.

Read-Only Memory (ROM): That part of the memory which contains instructions to the computer. It cannot normally be accessed or changed by the user.

Bit: A measure of the quantity of data stored or processed. One bit is one yes/no signal.

Byte: A measure of the quantity of data stored or processed. One byte equals a word consisting of eight bits, and is sufficient to describe a character or number. A *word* is the number of characters that the computer can process together.

Hardware: The combination of copper, steel, plastic, and silicon that makes up the physical part of a computer. Hardware alone cannot perform any functions.

Operating system: The programming which is built into a computer that enables it to function. The operating system "tells the hardware what to do," and how to process data with its memory, disks, keyboard, and other devices. The names of many operating systems contain the words "Disk Operating System" (DOS). It also tells the computer how to interface with the various items of software that will be run on it. The operating system is important because it defines the type of software that can be used, as well as the type of programming that can be done.

Program: A set of instructions for the computer to follow. A program defines the input, processing, and output that must be performed. Programs are written in various languages such as BASIC, C, Fortran, and APL in which the instructions are expressed in a way that can be interpreted by the operating system.

Software: A package of one or more programs which is designed for a specific application. Software will work with one or more operating systems and hardware configurations. General application software is sold as a stand-alone product for general use. Specific-application software may be written as a new program, adapted from an off-the-shelf program, or, in some cases, as a combination of both.

Bundled system: A computer system in which hardware and software are sold together, in one integrated package. In a bundled system, the hardware and software cannot be separated.

Unbundled system: A computer system in which hardware and software can be purchased separately. The software is said to be "machine portable" in that it will usually run on more than one type of hardware. One advantage of an unbundled system is that it enables the purchaser of the hardware-software package to use the hardware for other applications, or to purchase only the software.

Local Area Network: A means of linking together several computing components, such as a large mainframe computer, several terminals, several personal computers, printers, and plotters. The purpose of a local area network (LAN) is usually to allow users to communicate, share specialized equipment (such as a plotter), and share data. Networking can be accomplished with special hardwiring, use of telephone lines, satellite communication, or special networking packages including both software and hardware.

Terminals: Local workstations at which the system user can input data using a keyboard and receive output in the form of a visual display and/or in print. A "smart" terminal (usually a personal computer) is also capable of storing and processing data itself. It can, therefore, "upload" data that it has stored and/or processed to another computer or to other terminals, as well as "download" data which it receives from another computer. A "dumb" terminal is simply an input/output device and has no internal storage or processing capability.

Mainframe: A large, powerful, centralized computing facility supporting most if not all of an organization's data-processing needs. It is characterized by high-speed processing, extensive storage capability, multiple input and output facilities, a wide range of software, complexity of operation far beyond the skills of the average user, rigid environmental requirements, and very high cost. The organization with a mainframe computer will inevitably have a major organization function (Data Processing, Management Information Systems, etc.) that is responsible for servicing the various users. A mainframe is distinguished by its purpose, which is to provide massive calculation and storage capabilities to an entire organization.

Minicomputer: A powerful computing facility which can support all, some, or perhaps only one of the organization's data-processing needs. The minicomputer is still a sophisticated system, but it is more apt to be in the hands of the user rather than in a data-processing group. Most "minis" can function in any reasonable environment, are portable, and far less expensive than a mainframe. A minicomputer's (or workstation's) purpose is to support a small group of users with reasonably powerful computing capability.

Personal computer (PC): A small computer intended for one user. Because the PC is entirely in the hands of its user, it offers the benefits of convenience and responsiveness which larger, more complex systems cannot match. The very low cost of the PC not only makes it easy to implement, but also means that the risks of developing a system are small, as the worst that can happen is that we scrap it and try again. It is because of these features of low cost, user convenience, and ease of implementation that PCs are a good candidate for small-project applications.

Peripheral equipment: Devices that are external to the processing portion of the computer and provide input of data, output of information, storage, and special functions such as graphics.

Interactive system: A computer system in which the user interacts in real time with the computer. This is usually done with a terminal in which commands or questions are processed quickly and a result displayed. In comparison, the "batch" type of processing takes all the input at once and, later, delivers the printed output.

User-friendly system: A system which enables the user to interact easily with the computer, with simple commands. User-friendly systems typically feature Englishlike commands instead of the actual coded commands; "on-line tutorials," in which the system instructs the user as he goes; and helpful diagnostic messages when the user makes an input error. User-friendliness is extremely important in the small-project environment, since most users are not going to be skilled computer operators. The "graphic user interface," in which a "mouse" is used to issue commands and interact with images, provides a high degree of user-friendliness.

Menu-driven system: An interactive system in which the user is presented with a series of options from which to select, and then follows the instructions (or "prompts") as displayed on the screen. In effect, the computer is telling the user what to do. Menu-driven systems are relatively easy to learn and use and are appropriate for users with little system training. The disadvantage of the menu-driven system is that it is relatively inflexible, and this can frustrate the skilled user and prevent full utilization of the system's capabilities.

Command-driven system: An interactive system in which the user has a variety of commands which can be employed to tell the computer what to do. This type of system is more flexible than a menu-driven type, and enables the user to obtain exactly the results needed. However, it also requires more sophisticated computer knowledge on the user's part. Many systems are available that have both menu-driven and command-driven capability, and this is very useful in the small-project environment. For example, data from timesheets could be inputed by a clerical person with little training using a menu-driven routine, while decision analysis and cost and schedule forecasting were done by a project engineer using a command-driven routine. Often, a macro language can be used to create menu-drive applications for repetitive tasks within a command-driven system.

CONSIDERATIONS IN THE DESIGN OF PROJECT-MANAGEMENT SYSTEMS

In this discussion of project-management systems, it should be kept in mind that a computer cannot manage a project. What it can do is make the people responsible for the project as effective as possible.

As we saw in Part IV, the essence of project control boils down to the timely processing and presentation of useful information and was summarized as follows:

A project-management system is a means of presenting:

The right information
In the right format
To the right person
At the right time

The project-management functions that benefit from computer assistance include:

In the project-evaluation phase:
   Cost estimating and expenditure forecasting
   Planning and scheduling
   Resource scheduling
   Multiple-project analysis and scheduling
   Economic analysis
   Analysis of risk and contingency
   Bid preparation and analysis
   Management-level decision making
   Collection of project data
   Design and technical analysis
   Document control
   Contractor and bid evaluation
In the project-execution phase:
   Cost control
   Progress measurement and schedule control
   Productivity analysis and control
   Project-level decision making
   Materials management
   Control of labor and equipment resources
   Cost and schedule forecasting
   Contract administration
   Design decision-making
   Document control
   Quality control
   Risk management
   Field supervision
   Management reporting

To build a project-management system for a specific application, all we really have to do is combine the basic elements of system design to suit our requirements. These elements are:

Input in the form of keyboard entry, data from a storage disk, or communication from another computer.
Processing or data manipulation using equations and a programmed procedure.
Storage of data or results on a tape or disk.
Output in the form of displays, reports, and/or graphics.

As multiple project managers, all we really need to know about system design is how to define and combine these four elements of a system, such that the computer does exactly what we want it to do. Our system design becomes a specification that will be satisfied either by purchased software or by a program which we will have written, or a combination of both. The software will then be mounted on some hardware that always fits our specified requirements.

However, before a system can be designed, it is necessary to address some very fundamental questions about the work we do involving how we do it now, how we should do it, and how we could do it with computer assistance. In other words, the essence of system design is not programming, nor is it selecting the hardware and software. It is defining the job that the system is to do. To do that, we have to understand our own job very well.

BASIC STEPS IN SYSTEM DESIGN AND IMPLEMENTATION

To those of us involved in project management, it may seem like a lot of unnecessary work to follow all the system design steps shown below. After all, all of the system vendors claim that their product will solve all our project-management problems, so why not just go ahead and pick one? The number and diversity of available computer systems is, in fact, one of the reasons why the preparation of a system's specification is so important. The only way to cope with the volume of information, the pressure of time, and the intensity of salesmen is to have a clear idea of exactly what is wanted.

The system's design is also important because projects exist within an organization, and the power of the organization is mightier than any computer. Therefore, if our system is to be effective, and if it is to make our hard-won project-management methods effective, it must fit the organization and enhance its operation. To assure that this is the case, we must know ourselves, our jobs, and our needs—as well as those of the organization in which we work—before we can know what computer system will work for us.

The basic steps in system design are as follows:

1. Define user requirements
2. Develop system performance specification
3. Prepare plan and cost estimate for implementation
4. Obtain management approval to proceed
5. Prepare sample project and reports

6. Contact suppliers and develop a short list
7. Conduct "compute-off" and evaluate results
8. Obtain management approval for implementation
9. Implement system

Each of these steps is described in detail below.

Step 1: Defining User Requirements

This first step is both the most important and the most difficult. It requires a thorough analysis of current and future job practices as they relate to the receipt and manipulation of data, and the presentation of information. The "primary users" referred to below are the typical persons in the organization function for whom the system is designed (e.g., the project leader). There are, of course, other people in other functions who might also be users; we will refer to them as "secondary users." Taking the perspective of the primary user, we can develop the necessary definitions by considering the following questions.

Review and Definition of Existing Practice

Who will be the primary users of the system?

What are the objectives of the primary users' job functions? What methods are now used to perform these functions?

What information or data do the primary users get now? From whom do they get it, and in what form?

What information do the primary users provide to others? What information is provided to whom? To what use is it put?

What data manipulation is now performed by the primary users in order to provide the necessary information to others?

What problems exist with current practice?

What improvements could be made to the current practice?

Who would the secondary users be, to enhance current practice? What are their information needs? What information must they provide to others?

Definition of Ideal Practice

Can or should the primary users' job function be changed? If so, what should it be?

What methods should ideally be used?

What information do the primary users need to do their jobs ideally? From where should they get it, and in what form?

To what organizational functions should the primary users provide information? In what form should it be provided? To what use would it be put?

What data manipulation should be done by the primary users to provide this information?

Who would be the secondary users in the ideal situation? What information would they receive? What would they provide to others?

What improvements would the ideal practice achieve over current practice?

Specifying Short-Range System Requirements

What existing user functions could be computerized for greater efficiency?

What immediate improvements could be gained by computerization?

What existing computer systems might be used?

With what existing computer systems might the new system interface? What would be the type of data exchanged across the interface?

What level of computer capability will the primary users have? What other levels of user capability need to be accommodated?

What organizational functions will be putting in what data?

How much data needs to be stored? How much of that needs to be readily accessible for calculations and analysis?

How fast do the calculations have to be done?

What systems exist for coding project information? If new coding systems will be developed, what constraints (e.g., number of digits) may be imposed by existing practices?

How much flexibility is required in the system's operations? What operations need to be flexible and to what extent?

What cost limitations will apply to the design and implementation of a computer system? What schedule limitations apply?

Of all the specified features, what is the priority of each one? For example, is price more important than storage capability? Do we prefer buying a software package "off the shelf," or designing our own? Can we separate our "wants" from our "needs" in terms of system requirements?

Specifying Long-Range System Requirements

Specifying requirements requires a reevaluation of the issues outlined in our discussion of short-range requirements, but with a view toward the future. Any system we develop must be able to meet future as well as current requirements, or be adaptable to newer systems as they become available. So, for example, if the workload of the project-engineering department is expected to double in the next two years, we must be sure that our system is expandable to that capacity. The company may change or add to its computer systems, so any systems we introduce will have to be compatible with the new system. New codes of accounts, procedures, and interfaces may also be implemented.

One of the key aspects of system design is the identification and evaluation of "wants" versus "needs." Just as in any design function, there is a tendency to include capabilities that are desirable but not really necessary. The problems of distinguishing between wants and needs is particularly difficult in system design, since so many of the benefits have to do with "managing better" and are therefore difficult to quantify. One way to keep the system's specification under control is to specify the minimum system that will do the minimum job, and then put each added feature or capability through a cost/benefit analysis, just as one would do for a proposed design change.

Step 2: Developing System Performance Specification

The previous step was aimed at determining what we want the system to do for us. This step involves describing the specific characteristics of the system that will have the capabilities we need, both now and in the future. The system design consists simply of describing the four basic system elements: input, processing, storage, and output.

System Input Specification

In this step, we specify the type of input, the amount of data to be input, the sophistication of the person making the input, the location of the input function, the flexibility required, and the need for interaction with the main system during input.

The input to the system can take the form of interactive terminals, or electronic data transfer systems such as disks or tapes from other computers, or through electronic communication.

If input is to be in the form of interactive terminals which, in most cases, fits the small project environment, we have a number of options:

"Smart" terminals that are actually small microcomputers and provide local computing capability. Such terminals can process input data so that it goes into the main system in a certain format and the user can perform certain functions at the terminal without having to access the rest of the system. When the terminal performs some processing and then inputs the result to the main system, it is said to be "up-loading" information into the system. When the main system performs some processing and then sends the resulting information to a terminal, it is said to be "downloading."

"Dumb" terminals, which are strictly input/output devices for the main system.

"Local area networks," in which a number of computers and software packages are linked together and able to interact with each other. Most of the major manufacturers offer this capability.

Menu-driven input and/or command-driven input.

Remote terminals with tie-in to the main system via telephone, telegraph, or other forms of telecommunication.

An example of input by data transfer is the use of timesheet data that is already input to the accounting system. We might arrange for the work-hours in each labor category to be input to our cost control system.

System Processing Specification

In this step we must specify what the system must do with the input data: that is, what calculations are to be performed, how much data needs to be handled, how many calculations need to be made, how quickly the calculation must take place, what data files need to be read, what comparisons are necessary, and what results are desired. It must also be recognized that an interactive system allows the user to follow and even to affect the data processing function and we therefore need to identify those calculations that are always the same and require no interaction from those for which interaction is desirable. For example, the preparation of weekly expenditure reports would, in general, not be an interactive process, whereas the routines used for decision-making would be. Therefore we need to specify the operations which are to be interactive, and the extent of interaction desired.

The system processing functions that are most often used in project management are: arithmetic calculation, data retrieval, comparison, and extrapolation.

Calculation functions are performed the same way we would prepare a progress report, an estimate, or any other simple task. As we have seen in previous chapters, much of the work involved in progress-report preparation is the calculation of current figures and the comparison of those figures with the budget plan. To do this by computer, we retrieve the file of budget data and perform comparison of actual data vs. the budget plan.

We have also seen in previous chapters that extrapolation of current trends is a useful way to highlight potential problems and this is another function that is easily computerized.

These functions are generally provided by the purchased software. One feature to look for in software is the flexibility to define the functions that are to be performed. For example, there are many scheduling packages that do network analysis, but they all vary in terms of the user's flexibility in defining the coding system, to select arrow or precedence notation, to define subprojects, to load resource and cost data onto the activities, etc.

Another important feature to look for in software is the ability of one software package to interact with another. For example, in project management we are interested in network and resource analysis, cost estimating and forecasting, and management reporting. We may also wish to use other software categories such as cost estimating, risk analysis, spreadsheets, word processing, management graphics, and databases. Although most systems have software to do each of these functions, we may find that the ability to tie them all together will vary between systems. If these functions can be integrated a much more efficient system will result.

System Storage Specification

In this step we must specify the amount of data to be stored, the format for storage, and the requirements for recall. For example, we might wish to construct a database of norms for estimating, derived from past projects. This database would be carefully designed and probably integrated with an estimating routine. We might also wish to store design, cost, schedule and resource data from past projects for future analysis.

System Output Specification

The success of system implementation depends, to a great extent, on the effectiveness of the output. If a system produces clear, concise, useful, and easy-to-read charts and reports, it is bound to be a success. Conversely, many good systems fail to win the acceptance they deserve, because the users simply cannot relate well to the output produced. This principle is particularly useful in the multiple small-project environment where, as mentioned earlier, we are trying to present "the right information, in the right format, to the right person, at the right time."

There are three types of output that are relevant to multiple small projects: printed output, visual displays, and graphics.

Printed output   The reports generated by printed output will probably be circulated throughout the company to advise managers of status and problem areas. They should permit management by exception and provide only that information that is necessary to the particular report recipient. In order to achieve this, a system feature that is most useful is the flexible report writer. This feature enables the design of many reports, each of which has a separate format and that displays specific data. Systems without this feature have one or more report formats that cannot be changed.

Another useful feature in report preparation is the ability to sort, select, and order data. This makes it possible for us to select the specific information we want for each report, sort it into appropriate categories, and print it in a given order. For example, a manager's report listing all projects might be designed such that the project with the largest cost overrun is printed first, and subsequent projects ordered by decreasing overrun amounts. From our array of project data we might select cost data for an accounting report, and order it by cost code: the same data might be arranged by contract number for a report to the contracts engineer.

Visual displays   The image that an interactive system makes on the screen is, of course, only important to the user. However, visual displays are an important form of output because the user's effectiveness depends to a great extent on the system's ability to give quick outputs of useful information. One useful feature of some new systems is the ability to present several displays on one screen. Another feature to look for is the display of messages that tell the user what is going on when the computer is calculating, and also what is going wrong. Finally, we would like to have a visual display that makes it easy to see the results of a change to the project parameters. For example, we might be varying logic and durations to test the effect on the end date. We would, therefore, like a display format that makes it easy for us to see the end date after each run.

Graphics   Nowhere is the principle that "a picture is worth a thousand words" better illustrated than in management graphics. With well-designed graphics, the meaning of an otherwise confusing and uninteresting set of figures can be driven home clearly, quickly, and forcefully. And, as we saw in Part IV, a great deal of the effectiveness of our project-management effort will hang on our ability to focus the attention of busy managers to those areas in which it is required. So management graphics deserve consideration in the small-project environment. Most computer systems for project management have such a capability.

To specify the graphics capability required, we need to identify the type of graphics, the quality required, the source of data to be used, and the quantity required. For project management, the most common graphics types are:

Barcharts
Networks
Manpower histograms
Organization charts
Piecharts
Tracking and trending curves

The quality determines the type of drawing equipment required. For example, if top quality color graphics are desired, then a four- or eight-pen plotter with associated software is needed. This however, can add significantly to the system cost, although the cost of some plotters is now well below $1,000. Plotted management graphics are the best possible way to communicate plan and schedule information, and are strongly recommended. These may often be obtained by exporting project data to a spreadsheet or graphics program.

Step 3: Preparing the Plan and Cost Estimate for Implementation

At this point, we have an idea of what we are trying to accomplish with computerization, and what kind of system will be required. Before going further, it is appropriate to exercise some project-management techniques on ourselves, and establish a plan and estimate for what is to come.

The major cost items that can be expected are as follows:

Purchase or lease of hardware (i.e., computer and peripheral equipment)
Purchase or lease of software
Purchase of accessories
Maintenance contract for hardware and/or software (if applicable)
Modifications to office facilities to accommodate computer (if applicable)
Time required for training and implementation
Consulting services for applications, training and support (if applicable)
Supplies (computer paper, disks, etc.)

The computer acquisition and implementation should, in fact, be treated as a project just like any other, with a budget and a schedule. It should be recognized that the estimate prepared at this stage will be preliminary, in that we have not yet gone into the marketplace and obtained quotations. However, it is important to prepare a preliminary budget at this stage in order to accomplish the important next step: obtaining management approval to proceed.

Step 4: Obtaining Management Approval to Proceed

The purpose of this step is to advise management of what we are doing, obtain their general approval of the overall plan, and specific approval to proceed with system evaluation and recommendation. This step is suggested for three important reasons:

1. We wish to have company management "on our side" in this endeavor, since the organizational implications of a new computer system tend to be far greater than the capital cost might indicate. We want to be sure that management understands and supports the objectives of better project management, towards which the computer is merely a means. We can point out that the benefits of formal, multiple-small project methods and systems include: reduced project risks, increased consistency between projects, improved decision-making, greater accountability, more flexibility, increased productivity, cost-effectiveness, and greater responsiveness.

2. We wish to impress upon company management that the computer system is intended to be a cost-saving device, which is to be evaluated and managed like any other small project. Far too often, engineers and managers decide first that a computer is needed, then decide what kind of computer to buy, and never subject these subjective decisions to the same sort of scrutiny that any other investment decision would endure. We must never lose sight of the fact that there is only one good reason for computerization: that it will be profitable. We want our management to know that we are operating that way.

3. By providing preliminary cost/benefit figures, we are in a position to review our plans with management, and perhaps identify some changes. For example, we might be pleasantly surprised to find that management is enthusiastic about the project, and encourages us to specify and implement a more ambitious system than we would have done on our own. And, since small-project management affects so many parts of the organization, there may be implications and interfaces that only management are aware of, and that insight is an important part of a successful system.

Once we have obtained approval to proceed further, we are ready to journey into the world of modern computing. But before we are swept up in electronic excitement, we need to prepare ourselves for this journey. As always, the best way to be prepared is to know exactly what we want.

Step 5: Preparing the Sample Project and Reports

Although we went to a great deal of trouble in Steps 1 and 2 to define our needs, we had better expect that there will be a lot of systems that will meet or exceed them. We had also better expect that we are likely to see a bewildering array of computers, salesmen, and impressive demonstrations, at of the end of which we are likely to say, "But they all seem so good. How can I possibly choose one?"

The truth is that all the systems are probably very good. Each has features which its proponents claim make it immeasurably better than the others. The only difference between them will be that some will suit our needs better than others, and one will suit us best of all. But how will we know when the right one comes along?

The answer is to prepare a sample project (or projects) that represents the typical situation that our system will have to handle. We should also make up some input data in exactly the form in which we would normally receive it and design some management reports that show exactly what we would like the system to produce. Now, instead of approaching our friendly computer salesperson and saying, "Show me what your system can do," we can say, "This is what I need, show me that your system can do it!"

We have therefore defined system requirements, system performance specification, and a sample project for use in testing and evaluation. We know what we want and have expressed it in specific terms. At the same time, we are free to change any aspect of our system design as we become aware of new system features and capabilities. We have also established a basis for fair testing and comparison of the systems selected for in-depth evaluation.

Step 6: Contacting Suppliers and Developing Short-List

The purpose of this step is to identify three or four systems that are worthy of in-depth evaluation.

When conducting this "survey" of what is available, it is well to bear in mind these few hints:

1. *Consider software before hardware.* The effectiveness of a system will be determined first and foremost by the software: if the programs meet our needs, then it is likely to be relatively simple to arrange the hardware as necessary.

2. *Take the time to read the user's manual.* Short of running the software, there is no better way to get a feel for what the software can do than to read the user's manual. The clarity and completeness of the manual is itself a good indication of the quality of the software package.

3. *Avoid writing new software.* Many people jump too quickly to the conclusion that their needs can only be met by writing their own software. This may be true in some cases, but, in most project-management applications, there are few functions that cannot be satisfied by some existing software package. Although the existing package may represent a compromise, it must be remembered that the time, cost, and risk are greatly reduced. Software-development projects are notorious for overrunning budget and schedule by several orders of magnitude.

4. *It pays to be skeptical.* For example, one should pay no attention to claims that a certain capability we require will be available "on the next release of the software." We are interested (at least initially) only in that which is available and demonstrable now.

5. *Insist on talking with someone who is knowledgeable in the specific area of interest.* With the sudden proliferation of computer stores and companies, combined with the proliferation of hardware and software products, we could hardly expect an equal proliferation of experts. It is, of course, impossible even for a real expert to be familiar with every application of every system and, as a result, good advice is hard to find. So, when discussions progress to the point where it is possible to identify the items in the hardware or software product line which are of interest, it is worth the extra trouble to find and talk to the person who is most knowledgeable about those specific items.

6. *Examine closely any claims about hardware and software interfaces.* Experience has shown that the problems of getting machines to communicate with each other and the problems of linking software are often far greater than anticipated.

At the end of this step, we should have identified three or four systems for in-depth evaluation. More candidate systems will only add to the time and cost of the evaluation process, and will probably not increase the likelihood of the correct system being chosen. We may, during this step, also have made some revisions to our system design, to reflect what we learned. Now we are ready for the "try before you buy" part of the project.

Step 7: Conducting the "Compute-Off" and Evaluating the Result

The purpose of this step might be compared to selecting a new car by driving each of the cars selected for possible purchase around the same course and evaluating the results according to a set rating sheet. In this case, our "course" is the sample project, data, and reports.

Acquiring the Software

We begin by acquiring a full copy or demonstration copy of each of the candidate short-list programs. The low cost of personal computer software usually makes purchasing a viable route. We then learn each program, and use it to test its ability to plan and analyze the sample project. We then test its capabilities against our software specificiation.

Preparing the Rating Sheet

The categories on the rating sheet, and the relative weight given to each one, are determined by our definition of the system's design requirements. However, we can identify some system characteristics that will inevitably be important for a project application.

1. *User-friendly interaction*, i.e., the system should be easy to use, with a low "frustration factor." User-friendly features include English-type commands (as opposed to letters or numbers), tutorials (in which the system actually guides the user along), helpful error messages (e.g., "system will not accept decimal input" is more helpful than "syntax error") and help commands (in which, when in doubt, the user simply types "help" for a complete listing and explanation of the commands available at that point).
2. *Flexible reporting format*, or a fixed format that is well-suited to the particular application.
3. *Good interface characteristics* between the hardware and any existing hardware in the company. For example, it is a plus if our new microcomputer is capable of communicating with the existing mainframe computer even if we have no intention of doing so at present.
4. *Good interface characteristics between software packages.* For example, can we pass data and information from one software package to another?
5. *Good user support* from the hardware and software suppliers.
6. *Software documentation* providing a clear, easy-to-use user manual.
7. *The provision of training programs* for users.

8. *The ability to upgrade* hardware and/or software to increase capacity or to use newer products.

We should bear in mind that we are rating the software, the hardware, and the supplier in terms of price and performance.

Evaluation of the Results

To evaluate the results, we like to use a small committee of individuals with diverse skills, such as computing, project control, estimating, accounting, and management reporting. Each individual rates each system, and a composite score is agreed-upon by the committee. Evaluations should be made, at first, without regard to price. Once the performance ratings are established, then price can be considered and cost/benefit analyses performed.

Step 8: Obtaining Management Approval for Implementation

The purpose of this step is to present a specific proposal for implementation to company management, and to obtain approval for full implementation. As before, it is important to treat the subject of computerization objectively, and show in specific terms how profitability will be enhanced, due to time and cost savings on projects and in the project engineering and management functions. It is important that the first step in computerization be one which provides visible and significant benefits, so that the concept is well accepted.

Management support is even more important at this stage. During implementation, people are likely to react with cautious enthusiasm, tolerance, grudging acceptance, reluctance, and even outright hostility. We will be much better equipped to handle this if we, and everyone else, know that company management supports the program.

The proposal to management will probably be an update of the preliminary proposal, with greater details on the time, cost, and plan for short-range implementation, the possible longer-range developments, and a detailed economic justification. The organizational aspects of computerization will be uppermost in the minds of management, and that should also be squarely addressed by describing how the roles of existing functions may be changed, how lines of communication will be altered, and how efficiency in various departments will be improved.

Step 9: System Implementation

Setting up the First Project

After installation of our computer system, the next question is, "where to start?" Many companies find it useful to take a project and implement it on the new system while continuing to control it using the old methods. This provides a check on the results and avoids making a "guinea pig" out of the first projects. Often there are lessons learned from the early projects that indicate improvements that can be made before the system is implemented on a broad scale.

Keeping Everyone Informed

A key factor in the acceptance of the new system is that everyone must be kept informed, and we must display a willingness to share experiences with the new system and to help others obtain similar benefits. It is especially important to be sure that those individuals and organizational functions that interface with our new system be made to feel that they were consulted early enough to assure that their needs and constraints were considered.

Some of the methods for keeping people informed include progress memos, coordination meetings, and in-house seminars and training sessions.

Preparation of Procedures

It is important to assure that the new system is used in a manner that is consistent with the rest of the project's engineering and management functions. The best way to assure this is to prepare procedures that specify how data is input, processed, and output. The procedures should show how each organizational function interfaces with the system, as well as specify who is in charge of the system. Each function that will use the system should have specific procedures to follow.

Many systems have different levels of security, which are protected by passwords and other procedures, and the limits of each user therefore must be defined. For example, payroll or personnel data can be included in the system, but access to that data must be restricted. Similarly, we might construct an estimating database that is available to any project engineer for estimating work, but which can only be updated by those responsible for it. In that case, access to "read" from the database is unlimited, but access to "write" to the database would be restricted.

The procedure should also specify who the "computer czar" is to be, that is, the person who answers questions, resolves problems, sets priorities, listens to complaints, deals with the vendors, and generally coordinates system use. The recent surge in the use of personal computers in business has had a somewhat undesirable side effect of diluting the supervision and coordination of computer usage within a company. In past years, companies purchased a mainframe computer to service the entire organization, and users had to obtain the services they needed from the Data Processing (D.P.) Department. Because systems were expensive, maximum utilization was the goal, and users were often faced with a long wait for the needed capability to be provided by an already overworked system. In recent years, the microcomputer has enabled the user to implement his own solution in his own way, without recourse to the D.P. Department. While this has broadened the effectiveness of computers, it also has created many situations in which individual effectiveness has improved, but organizational efficiency has not. Therefore, today's responsible user has to assume some of the responsibility previously borne by the D.P. Department, to make sure that others are aware of what they are doing and that their system is compatible with others in the organization.

Training

Formal training courses are a significant aid in the acceptance and effective utilization of the system. The training course can be based around the user's manual, which describes how to operate the system, and the procedures, which describe how the system fits into company operations.

A REVIEW OF PROJECT-MANAGEMENT SOFTWARE

Since most project-management needs can be satisfied by a variety of existing systems, it is worthwhile to review some of the characteristics of current systems. Project-management software can be divided into a number of categories as shown below.

Planning and Scheduling

A great many software packages are currently available to perform planning and scheduling. Virtually all of these perform forward and backward pass calculations of critical path and float. Some of the features that vary between software packages are:

The size of the project handled (i.e., number of activities)
The ability to use precedence or arrow notation

The flexibility of the coding system used to identify activities

The ability to create and use subprojects

The ability to calculate free float

The ability to perform resource allocation, resource aggregation, and resource leveling

The number, type, and flexibility of resources permitted

The flexibility of displays and graphic output

The speed of analysis

The ability to handle multiple projects simultaneously

The ease and flexibility of updating (e.g., out-of-sequence updating that allows updates even if activities have not been progressed in strict logical sequence)

The number of users permitted

The amount and flexibility of data that can be attached to each activity

The selection of a package will depend, of course, on the needs and priorities of the user. These features and the selection process are discussed more fully in the next chapter.

Cost-Estimating Software

It is interesting to note that far fewer cost-estimating packages are available than scheduling packages. The reason for this could be due to the fact that the approach taken to cost estimating is less standardized, and it is therefore more difficult to create a standard package that would satisfy many users.

There are a number of software packages available that are designed for specific types of estimating, such as architectural work or process industry projects. If an available program suits the particular application, all well and good. However, there may not be a program that exactly fits the user's application. Fortunately, there are a number of software packages available that are suitable for general estimating applications. If we consider that cost estimating usually consists of a lot of multiplication of quantities by unit costs, and addition of the results, it is evident that it is a simple matter to computerize it. The methods and data required for estimating can be developed with available software as well. Some of the types of programs that can be used are as follows.

"Spreadsheet" Software

Spreadsheets enable the user to define a matrix of data as well as the calculations required to define additional data. One can, for example, take a column of quantities and multiply it by another column of unit costs to create a detailed estimate. One can set up spreadsheets for detailed calculations that are then summarized and subjected to further manipulation in a higher level spreadsheet.

Database Software

Cost-estimating applications are designed to collect, analyze, normalize (i.e., adjust to a consistent basis or "norm"), and present the basic data needed for estimating. Since the best predictor of future costs is usually our past experience, and since small projects continuously generate a lot of useful, actual cost data, a database is a handy way to organize it all.

Databases and spreadsheets can be integrated with planning software in order to automate the estimating process described in Chapter 5, in which the labor and construction equipment resources are estimated from a standard file of rates. This makes it possible to examine the cost impact of different planning scenarios, as well as to make cost forecasts reflecting current progress.

A cost-estimating database should have an independent variable (e.g., design parameters such as cubic yards of concrete), and a corresponding value for the dependent variable (cost). It should be organized according to cost code, such that data can be easily selected, sorted, and summarized. Other database applications are discussed below.

Economic-Analysis Software

There are a number of software packages available that perform economic analysis in order to evaluate project profitability. These programs can be useful in establishing the return on investment, or in monitoring the projected return as variations to the original basis occur. It is worth noting that updates in the profitability analysis are seldom done after the project is approved, although they should be, as it is a relatively simple task for the project engineer to use an economic analysis to make decisions based not on minimum cost or schedule, but on maximum profitability.

Database Software

There are a number of database packages available, some of which are intended for use in conjunction with a network package. The database, as can be seen from the cost-engineering applications described above, enables us to collect, organize, analyze, sort, select, compare, and present any kind of data in any way we want. Some of the project-management applications of databases, in addition to estimating, include:

Data on the durations of typical design and construction activities

Material control by storing and accessing data on purchase orders, purchase requisitions, delivery dates, quantities, etc.

Document control (i.e., keeping track of drawings, specifications, and critical communications)

Vendor data

Contractor performance data

Maintenance data (i.e., required inspection and maintenance intervals and procedures, and actual maintenance records for each item)

Weight data (for those projects in which weight is critical)

Personnel data

Wage data

Some database programs are like "blank slates" and can be designed in any way that the user requires, while others are designed for a specific application.

A very useful feature of some systems is the integration of a database with the network-analysis program. This feature enables us to identify a resource requirement for an activity, and link that resource to a detailed set of relevant data in the database. For example, we might identify heat exchanger E103 as a resource requirement for the activity "erect heat exchanger." By setting up a link to our database we can have the program select some or all of the data pertaining to E103 and advise us of such facts as the vendor, the purchase order number, the price, the promised delivery date, the current status, the current delivery date, etc. We can even use this capability to update our schedule automatically to reflect current deliveries, or to update our cost forecast automatically to reflect current progress.

Management-Reporting Software

Management-reporting software consists of word processing, special report-generating packages, and management graphics. This software may be available within the existing computer system of the company. Features to look for in these packages include the ease of use, compatibility with existing systems (such as word processing), and interface characteristics with the main project-management programs. It is highly desirable that the report-writing and management graphics packages either be part of, or fully compatible with, the planning, scheduling, cost estimating, and project-control package(s) to avoid duplication of input and assure timely reporting.

Risk-Analysis Software

For projects involving significant financial risks, a risk-analysis program may be helpful in addressing the probability of various costs and schedules being achieved. For example, bidding strategy may neccesitate an evaluation of bid price vs. the probability of getting the job. Or a turn-around project that incurs major costs for every day the unit is shut down may benefit from a probabilistic analysis of schedule to determine the probability that the unit will be running on a given date, and to be able to plan accordingly.

Probabilistic analysis is appropriate for many more project-management situations than is generally acknowledged, probably because many engineers and managers are somewhat suspicious of statistics as a means of decision making. However, there are a number of software packages available that do a credible job of analyzing the probability of cost over-run or schedule delay (see Chapter 6). In general, these programs evaluate costs separately from schedules, and use a Monte Carlo simulation to generate the results. In Monte Carlo simulation, a random-number generator is used to select a value for each activity duration or cost variable, until a sample cost or schedule outcome is calculated. By repeating this process hundreds or even thousands of times, a probability function describing all possible outcomes can be developed. The major drawback to this type of analysis is that it requires that the variables be independent, which they seldom are. However, most programs provide a method to adjust for this.

If risk and contingency analysis is important, an appropriate software package can make the inevitable judgements more rational, credible, and acceptable. And, an appropriate package can point out the magnitude of risks that were previously thought to be inconsequential.

CURRENT TRENDS IN PROJECT-MANAGEMENT COMPUTER SYSTEMS

There are two basic trends observable today relevant to those in the project-management field:

1. New hardware and software products and technology that can be used effectively in the multiple small-project management field are being introduced and being implemented constantly. New programs often provide generous features specifically to assist with multiple projects.

2. The recognition of project management as a business endeavor that requires computer assistance is also increasing.

As a result it is now considerably easier to select, justify, and implement a project-management system than it was a short time ago. In addition, these trends are continuing and even strengthening today. Some of the new products on the market or approaching the market include:

Graphical user interface: These systems use a small, handheld device known as a "mouse" to move the cursor on the screen and thereby input commands simply by pointing to a display of command options. The keyboard is required only for the entry of data. The user interacts with the software in a graphical way, rather than by typing characters. This technology is quite useful since graphic images (e.g., network, bar-chart) are often used in project management.

"Windowed" displays: These are systems that are capable of displaying the outputs of several programs in several "windows" at once. For example, a project engineer could update his network and database, while preparing a tracking curve, and see the results of these operations simultaneously on the screen. Or, when preparing his progress report, he could use the word-processing, graphics, and spreadsheet programs in an integrated mode to perform a calculation on the spreadsheet, to plot the results, and to "paste" it into a report.

Integrated software: The ability of a program to do many related functions, and/or to explore or import files from other software, has been greatly improved and is continually being made easier.

Local Area Networks (LAN): These local networks allow the various computers, terminals, workstations, and central processing equipment to communicate directly, thereby giving the user the power to utilize the entire office system from his workstation. An example of LAN is electronic mail, in which a memo or report prepared at one workstation can be transmitted electronically to other workstations as well as be printed for distribution. LANs are also used to link large and small computers to make the added power of the larger machine available to the micro-user. And specialized hardware, such as a printer, hard disk, or plotter, can be utilized from several different workstations. The LAN has relevance to multiple-project applications when it is necessary to view or retrieve files on other projects planned by other people.

Machine speed and capacity: The power and speed of minis and micros is rapidly increasing, and this extra power is often used to improve ease of use. That is especially helpful in the multiple small-project environment, where little time is available for project management tasks.

Organizational changes: As systems become easier to learn and use, as well as more effective, their use is extending in all directions in the organization. Managers who, a year ago, would have never thought of using a computer themselves now rely on their own desktop model. The same is true for many people at the working level. One notable result of this is that the organizational structure for control of the computing function is changing from the data-processing department (designed to fit the mainframe computer) to a more integrated function.

COMPUTER-AIDED DESIGN APPLICATIONS FOR PROJECT MANAGEMENT

During the past several years, the trends toward greater computing power at much less cost have been very evident in the area of Computer-Aided Design (CAD). As a result, both owner and contractor companies in many fields are finding that CAD offers great improvements in the efficiency with which design tasks are conducted. Therefore, project engineers and managers are becoming increasingly aware of and involved with CAD as a design and drafting tool. However, CAD systems also offer the potential for dramatic improvements in the project-management function as well. CAD systems are particularly relevant to small projects in a large operating plant that has an engineering fucntion of sufficient size to warrant CAD systems.

Basic CAD Components and Capabilities

A CAD system is designed to perform many of the functions associated with the design process, including:

Drafting, including layouts, revisions, and final drawing
Storage, recall, and display of relevant design information
Material takeoffs for procurement
Modeling in three dimensions to remove interferences
Performance of design calculations
Application of design standards

To accomplish these tasks, a CAD system typically consists of the following components:

A workstation consisting of one or more screen displays, an electronic table, a keyboard, and a device for locating points on the screen or

on the table (e.g., "mouse"-type cursors, light pens, "joysticks," thumbwheels, etc.)
A central processing unit providing the hardware and software to perform the necessary operations
Storage facilities such as tape or disk
Auxiliary input devices (as required)
Telecommunications facilities (as required)
Printers and plotters to prepare the drawings

In most cases, a single processing unit can support a number of workstations, which also share the output devices.

As a result of the efficiencies introduced by CAD, both owner and contractor companies have experienced improvements in design efficiency by a factor of four or more. CAD systems are of interest, not only because their use is becoming widespread, but also because they have some features that are of great interest to the project engineering and management functions. Foremost among these features is the database capability that enables us to attach management data to design graphics. Current technology also embraces Computer-Aided Manufacturing (CAM), in which the design of a part to be manufactured can be translated directly to a tape for a numerically-controlled production machine.

CAD Applications to Cost Control

One of the most common and long-standing challenges in cost engineering involves the exertion of cost control during the design phase. Typically, this effort takes the form of procedures to control design man-hours and costs, as well as to estimate and track design changes. However, the most important aspect is usually not addressed, and that is control of the cost implications of the many design decisions that are made. This control is vital, because the design of the project is the aspect over which we have the most control, and it is also the aspect that has the most potential impact on cost.

Unfortunately, most companies are not very successful at integrating design and cost engineering. This is partly because of the misplaced belief that engineering work should be done in the minimum time and cost. This, of course, discourages any discretionary work such as efforts to reduce cost through design changes. Another problem is that most design engineers have neither the time, data, nor inclination to get involved in cost studies, just as most cost engineers are not capable of design work.

The CAD system enables us to construct a database for estimating the cost of each of the components included in the design. Thus the cost of each alternative design configuration can be easily calculated and provided to the design engineer. Changes can also be identified and estimated more efficiently. Life-cycle costs can also be integrated into the design process in this way.

CAD Applications to Weight Control

Many projects, such as offshore platforms, require monitoring and control of weights. This can be done in a manner similar to that used for cost control, by constructing a database of weight information attached to the graphical representation of each design component. The resulting data and analysis of weights provides vital information for platform structural analysis and planning of construction operations.

CAD Applications to Document Control

All projects have the problem of storage and organization of documents. This is required not only for efficiency, but also for certification and operation of the facility. Since many documents relate specifically to a design component, the CAD system can be used to log in all documentation relating to a specific equipment or bulk-material item. Typical documents stored in this way include:

Quality control documentation, such as testing and inspection results
Operating information
Vendor information
Maintenance information

CAD Applications to Maintenance

A typical problem of engineering and construction projects, large or small, is the problem of "handling over" the completed facility to those who will operate it. The operation personnel are interested in data that was (presumably) accumulated during the engineering phase, but was of little interest to those who were designing and building the facility. Consequently, the operations staff often find it difficult to obtain and organize the information they need. One example of this problem is maintenance.

In order to assure that an effective program for inspection and maintenance is set up, we can use the CAD database to accumulate information relating to inspection and maintenance of each item shown on a drawing. The database could show, for example:

The frequency of inspection
The intervals between planned maintenance
The procedures to be followed for planned and unplanned maintenance

Once the facilities are in operation, and the maintenance program is in effect, additional information could be added to the database to facilitate planning and monitoring of the inspection and maintenance function. Additional information that could be added includes, for example:

The date of last inspection
The result of last inspection
The date for the next inspection
The date of the last planned maintenance
The maintenance performed
The date of the next scheduled maintenance
The maintenance to be performed

CAD Applications to Construction Planning

Most CAD systems have the capability to work in three dimensions. This capability makes it possible to model the construction work to be done and see potential interferences, misfits, or impractical construction requirements. For example, one of the uses of a plastic model is to be able to actually see, in three dimensions, what is going to be built. The three-dimensional model on a CAD system not only lets us see the project in three dimensions, but also enables us to "move things around" with ease. This feature can be used to test the feasibility of different arrangements, and to actually simulate construction operations such as major lifts. For small projects in dense operating plants, the three-dimensional modeling feature can be a big help in avoiding expensive situations in which the design looks good on paper but proves to be impossible to construct in the field.

Finally, by simple speeding up the design process and the production of approved drawings for construction, the CAD system can do a lot to make construction go more smoothly.

CHAPTER SUMMARY

In this chapter we reviewed the availability of computer tools for project-management applications. We saw that there are many computer systems that are appropriate for small projects. We also discussed the necessary steps for system design, specification, selection, and implementation. The next chapter will discuss the specific ways in which personal computers (PC's) can be used in the multiple small-project environment.

15

Using a PC for Managing Multiple Small Projects: Guidelines for Software Selection and Implementation

PERSONAL COMPUTERS: AN IDEAL TOOL FOR MULTIPLE SMALL PROJECTS

Advantages of Personal Computing

Personal computing might be defined as computing done by one person for their own benefit. If we accept this definition, we can see that personal computers (PCs) are ideal for the small project leader.

Small projects require project management tools that are easy to learn and use, and which produce useful results quickly. They are apt to require a low-cost, easy to implement solution (e.g., software with features that are specifically designed for multiple projects).

Personal computer hardware and software meet all of these requirements, and therefore, have found wide acceptance in the multiple project environment. Useful PC software includes:

Project Management (e.g., planning, scheduling, resource scheduling, updating) Typical software is described below.

Spreadsheet (e.g., cost estimating, status reporting, PERT analysis) Typical software includes Lotus 123, Quattro Pro, Excel.

Database (e.g., project data, vendor data, cost data) Typical software
　　includes Paradox, Q&A, DBase.
Graphics (e.g., management reporting) Typical software includes Har-
　　vard Graphics, Freelance, Powerpoint.
Assignment Modeling (e.g., used for assignment scheduling as described
　　in Chapter 8) Typical software includes Sagacity.

Planning should always be done by those who are closest to the work,
and, ideally, by those who will actually do the work. Thus, small-project
planning should be pushed down towards the workface, and out into the
organization as broadly as possible. The low cost and ease of use of PCs
make this practical.

A Broad Spectrum of Project Management Software

Project management software for personal computers is usually divided
into three categories: "low end," "high end," and "mid-range" (see
Figure 15.1). These categories, each of which provides useful features for
multiple small projects, are described as follows:

SPECTRUM OF
PROJECT MANAGEMENT SOFTWARE

"LOW-END"	"MID-RANGE"	"HIGH-END"
PART-TIME USER	PART-TIME USER	FULL-TIME USER
NON-SPECIALIST	SOME SPECIALIZATION	SPECIALIST
HIGHLY INTERACTIVE	ALLOWS USE OF MORE	GOOD FOR LARGE PROJECT
GOOD FOR SMALL PROJECT	ADVANCED METHODS	MORE FEATURES
LOW COST	MODERATE COST	HIGHER COST
EXAMPLES:	EXAMPLES:	EXAMPLES:
MICROSOFT PROJECT	PROJECT WORKBENCH	OPEN PLAN
TIME LINE	PERTMASTER	PRIMAVERA
SUPERPROJECT		ARTEMIS

Figure 15.1   Project Management software ranges from the low-end, for
part-time, non-specialist users, to the high end, for full-time planning specialists.

Low-End Project Management Software

This type of software is intended for a part-time user, who is not a specialist in planning, scheduling or project management. This means that the program is relatively easy to learn and use.

Low-end software is usually designed such that a plan can be created directly on the screen without the necessity of thinking or drawing it out beforehand. The "outlining" feature, in which a plan is derived from the user's input of outline-style information, is a good example of this capability.

Low-end software will usually provide surprisingly powerful and sophisticated features for single- and multiple-project planning. Therefore, they often represent an excellent choice for the multiple small-project manager. Typical programs in the low-end category include Microsoft Project for Windows, Time Line, and Superproject Expert. The cost of low-end software is usually less than $800.

High-End Project Management Software

This type of software is intended for a full-time user who is a specialist in project planning, scheduling and management. They provide, on a PC, the type of processing power that was previously available only on mainframe computers.

High-end software of course, provides the capability to plan and control large projects, with plans consisting of thousands of activities. They also offer more sophisticated functions and capabilities than low-end software. Formal training classes are often recommended for users of high-end software. Typical programs in the high-end category are Open Plan, Primavera Project Planner, and Artemis. The cost of high-end software is usually more than $2500.

Mid-Range Project Management Software

As might be expected, mid-range software is between the low- and high-end categories in price, user sophistication, and capability. This type of software provides additional features but also may require more skill on the part of the user. Typical programs in the mid-range include Project Workbench and ViewPoint.

Links Between Low- and High-End Project Management Programs

As PC project management software has matured, the ability to transfer data from low-end to high-end software has markedly improved. This can be a very useful feature in the multiple project environment (see Figure 15.2).

In some applications, low-end software is used to prepare the plan for each small project. Each project leader uses the fast, easy, friendly interface of the low-end program to prepare the plan, schedule, and resource requirements of their projects. This information can then be exported to a high-end program that a specialist planner uses to accumulate data from all the small projects and prepare a large multiple-project plan. In some cases, data from the high-end program can be exported back to the low-end program after it has been processed.

All of the low-end programs mentioned above have exporting capability to one or more high-end programs.

Links Between Project Management and General Purpose Software

An extremely useful feature found in PC project management software is the capability to export data to a general purpose program such as a spreadsheet (see Figure 15.3). Why might we want to do this? Some examples:

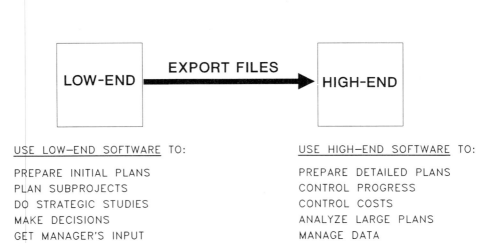

USE LOW−END SOFTWARE TO:

PREPARE INITIAL PLANS
PLAN SUBPROJECTS
DO STRATEGIC STUDIES
MAKE DECISIONS
GET MANAGER'S INPUT
PLAN EACH SMALL PROJECT

USE HIGH−END SOFTWARE TO:

PREPARE DETAILED PLANS
CONTROL PROGRESS
CONTROL COSTS
ANALYZE LARGE PLANS
MANAGE DATA
PREPARE MULTI−PROJECT
    PLANS

Figure 15.2   It is often useful to import and export files between low-end and high-end project management software.

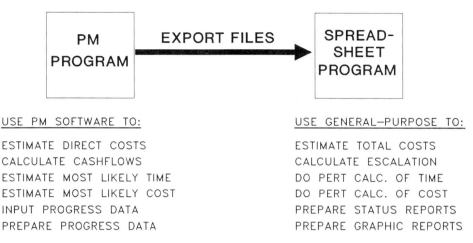

Figure 15.3 It is often useful to export files between project management and general purpose software.

Cost Estimating

The direct material and labor costs are estimated using the project management program, which also prepares an expenditure forecast showing workhours and costs per month. This information is then exported to a spreadsheet that completes the cost estimate by adding company overheads and other charges, and presenting it in an approved budget format. The spreadsheet program can also be used to prepare management graphics such as a pie-chart showing the breakdown of costs into various categories.

Probabilistic Analysis

The PERT calculation described in Chapters 3 and 6 can be performed nicely by a spreadsheet application. A project leader could prepare the plan and cost estimate using project management software, then export the appropriate data to a spreadsheet in which the PERT analysis is performed.

For example, from the project management program we might export a table showing the most likely duration of each activity. In the spreadsheet, we ask the user to rate project risks, resulting in an automatic calculation of maximum, minimum and 50/50 values. In this way, the schedule contingency required by each activity is determined.

For cost analysis, we might export, from the project management program, a table showing the most likely cost for each resource category. In the spreadsheet, a PERT calculation is performed to show the cost contingency required by each resource category, and, of course, by the project as a whole.

Progress and Performance Calculations

Project management software might be used to update the project model in order to show earned progress and workhours. This information can then be exported to a spreadsheet in which tracking curves and detailed performance analysis charts can be prepared.

Status Reporting

Status data from project management software can be exported to a spreadsheet from which highly customized management reports and graphics can be prepared.

SPECIAL MULTIPLE-PROJECT FEATURES IN PC SOFTWARE

As personal computers proliferated, software vendors apparently realized that many PC users could benefit from project management software. Many of these users manage multiple small projects, but do not consider themselves "project managers" in the traditional sense. Therefore, the vendors saw a market in these nontraditional multiple-project managers and have worked hard to provide software with features that accommodate the special needs of multiple small projects. Some of the more important and useful features are described below.

Outlining

Using Outlining to Plan the Single Project

The outlining feature, found in some low-end programs, is an extremely important advance in planning for small projects. While many people have difficulty adapting to the idea of a network diagram as a metaphor for a plan, everyone knows how to make an outline. So outlining has the great benefit of making it much easier for nonspecialist planners to become comfortable and efficient in creating CPM plans.

To plan with an outline, the small project leader might start by typing in the main headings of the outline (see Figure 15.4a). These main headings might be:

PLANNING WITH AN OUTLINER

PROJECT SUMMARY
   PHASE I: PRE-APPROVAL

   PHASE II: DESIGN

   PHASE III: INSTALLATION

   PHASE IV: STARTUP

Figure 15.4a   A first step in planning with the outline technique is to define headings.

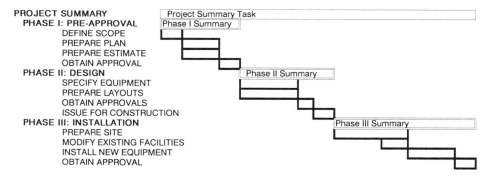

Figure 15.4b   Subsequent steps in planning with the outline technique involve adding subtasks, durations, and dependencies. Note that each summary task automatically rolls up the duration, resource and cost information from the subtasks below it.

Phases of the project
Major areas of operation
Work by different specialties
Major systems

Once the headings are defined, activities can be created within each heading (see Figure 15.4b). For each activity, duration, resource requirements and predecessors are defined. The time, cost and resource information

PLANNING WITH AN OUTLINER (Details Hidden)

PROJECT SUMMARY	Project Summary Task
PHASE I: PRE-APPROVAL	Phase I Summary
PHASE II: DESIGN	Phase II Summary
PHASE III: INSTALLATION	Phase III Summary

Figure 15.4c   An outline can be "collapsed" to show only the summary tasks.

from these activities will be automatically summarized (i.e., "rolled-up") by the heading.

Note that most outliners allow the heading and task data to be input in any order that they occur to the planner. Rearranging the outline, moving headings and activities up and down, or in or out, is quick and easy. Therefore, outlining makes it easy to plan a project without using pencil and paper first, but simply to do the creative planning right on the screen of the PC.

Finally, outlining usually provides the capability to present the plan and schedule at a summary level by concealing the activities and displaying only the headings (see Figure 15.4c).

All of these features make outlining a very important feature in the multiple small-project environment, where the time available for planning is at a premium, and where planning is often done by nonspecialists.

Using Outlining to Prepare the Multiple-Project Schedule

Outlining is also useful when small projects are being combined into a multiple-project master schedule. Projects can be summarized, moved around, and prioritized easily using the same outlining features.

Standard Resource and Cost Lists

Multiple projects often use the same "pool" of resources, as well as the same types of equipment and materials. To prepare a project model, of course, requires a list of resources and material costs. Therefore, many programs allow us to define a standard list of resources and costs, and to use that list over and over as each new project plan is prepared. This saves the time required to create the list, and also provides the consistency and uniformity required by multiple project analysis.

Multiple Calendars

When we consider the pivotal role played by the calendar in converting a plan to a schedule, it is evident why multiple calendars can be important when scheduling multiple small projects.

Imagine a small project designed to improve a manufacturing facility by adding a new type of gauge. Some of the engineering work might be done at the plant, where the engineers work from 8:00 a.m. to 5:00 p.m., and have 10 paid holidays per year. Detailed design work might be done by a design consulting firm, whose work hours are 7:30 a.m. to 5:30 p.m. (including one hour/day overtime) and who have eight paid holidays per year. Installation work might be done by a construction contractor whose forces are working 10 hours a day, 6 days per week. And the final hookup, testing and calibration might be done by the plant's own maintenance people who are available 24 hours/day, 7 days a week. This simple project therefore has four calendars, one for each category of resource: plant engineer, consulting engineer, construction labor, plant maintenance.

Imagine now a functional manager whose resource categories are individuals: John Smith, Sandra Jones, Robert Johnson, and John Robertson. Each of these resources (i.e., individuals) has a calendar that defines the days they are unavailable due to vacation, jury duty, sick leave and so on. In this case, the number of calendars per project is defined by the number of individuals working on the project.

Multiple calendars allow the scheduling calculation to deal with this. In many low-end programs, a calendar can be attached to a resource. The schedule calculation then looks at the resources required by an activity and *schedules the work only when all the resources will be available, according to their individual calendars* (see Figure 15.5).

High-end software may also offer the option of defining a number of project calendars, and indicating which of these calendars is followed by each activity.

Complex Calendars

Since calendars often must be flexible to accurately reflect the realities of the multiple-project environment, many project management programs, even at the low end, provide useful complex-calendar functions. A good program will permit us to define calendars with great specificity—defining the specific hours worked by a specific resource on a specific day. For example, we could specify that Jane Jackson will be available to work Saturday, June 10, from 12:00 to 5:00 p.m.

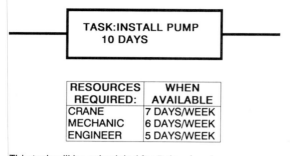

This task will be scheduled for 5 days/week.
The program will only schedule work when all resources,
according to their calendars, are available.

Figure 15.5   Each resource assigned to a task can have a different calendar. Tasks will be scheduled only when all resources are available.

Other useful calendar functions are those that permit global commands to change the calendar. For example, we might wish to specify that all Saturdays are working days, with hours from 10:00 a.m. to 2:00 p.m. With the global commands provided by some programs, this can be done with one operation.

Another useful calendar function is the ability to distinguish between shifts. In some cases, we can identify that a particular resource works a particular shift, and the shift calendar defines that resource's hours. The ability to assign a premium for shift and overtime work is also useful.

Multiple-Project Summarization Techniques

An important technique for analyzing multiple projects is the summarizing of subproject information into the master, multiple-project schedule (see Chapter 7). In this technique, a complete subproject is represented by one activity that summarizes the subproject information. Different programs will have different ways of performing this function. Some useful features for summarizing subprojects are:

The ability to move in and out of subprojects
The ability to update subprojects for schedule changes made as a result of multiple project scheduling
The ability to update the multiple project schedule for changes made in the subproject

The ability to identify that the subproject is using resources from the resource pool

The ability to view histograms shown in the summary activity that preserve the details of the subproject's resource utilization

The ability to prioritize and level summary activities

The ability to present the workload of each resource across multiple projects

Multiple-Project Combination Techniques

The second important technique for analyzing multiple projects is the summarizing of subproject information into the master, multiple-project schedule (see Chapter 7). In this technique, a complete subproject is incorporated into the master, multiple-project schedule with all of its activities intact. Different programs will have different ways of performing this function. Some useful features for combining subprojects are:

The ability to define constraints between activities in different subprojects

The ability to identify the project to which each subproject belongs

The ability to identify the workload of each resource across multiple projects

The ability to incorporate subprojects that have a different calendar from the master multiple-project schedule

The ability to set priorities for an entire subproject or for any individual activity within the subproject

The ability to update subprojects for schedule changes made as a result of multiple project scheduling

Choice of Scheduling Methods

Duration-driven

Many people find that the easiest way for them to plan is to estimate activity durations and resource requirements, and let the program calculate the work-hours of effort. This is called "duration-driven" scheduling (see Fig. 15.6a).

For example, we might plan a research task by saying it will take 2 technicians, 3 days to do it. The program will then calculate the work effort as 2 people × 8 hours/day × 3 days = 48 workhours of effort. Most project management software offers this approach to planning.

Effort-driven

Many other people approach planning by saying: "This task will take 160 workhours of effort." They find it easiest to input the effort and the number

INPUT: TASK DURATION
          NO. RESOURCES

OUTPUT: WORKHOURS OF EFFORT

ASSIGNMENT DURATIONS: SAME AS TASK DURATION

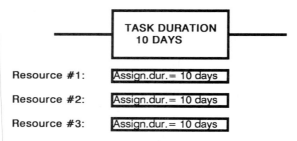

Figure 15.6a   Duration-driven scheduling assumes that resources are used uniformly over the duration of the task, and calculates the effort to be expended.

INPUT: WORKHOURS OF EFFORT
          NO. RESOURCES
          NOMINAL TASK DURATION (OPTIONAL)

OUTPUT: TASK & ASSIGNMENT DURATIONS

ASSIGNMENT DURATIONS: MAY BE DIFFERENT

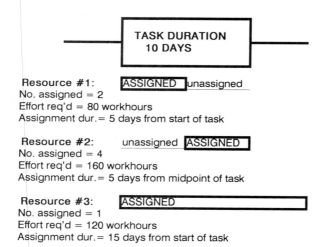

Figure 15.6b   Effort-driven scheduling requires the effort by each resource to be input, but allows assignment durations to vary.

of resources to be applied, and to have the program calculate the duration (see Figure 15.6b).

For example, we might plan a programming task by saying it will take 160 workhours of effort, and that we will assign two programmers to work on it. The program will then calculate the duration as 160 workhours/2 people × 8 hours/day = 10 days duration. Many programs offer this as an alternative approach to planning.

One advantage of having both types of scheduling available is that a project leader can first use duration-driven scheduling, to have the program calculate the effort. Then she can change the number of resources assigned, and see the resulting change in the activity duration. This is helpful in evaluating methods for schedule acceleration.

Another advantage of effort-driven scheduling is that, in some software, it permits the duration of a resource's assignment to be longer or shorter than the activity duration. For example, an activity may be scheduled for 10 days, but the resource "electrical engineer" may have only 16 workhours of effort to perform, and will be done in two days. This technique can have the effect of keeping the number of activities in the plan to a minimum, while preserving precision in the scheduling of resources.

Use of Priority for Multiproject Resource Scheduling

In the multiple project environment, a master plan for many small projects is created. When many small projects make up the multiple project schedule, the critical path has considerably less significance than it does for a single project.

Imagine a monthly maintenance plan, as shown in Figure 15.7. Each "work order" takes about three days to complete and is independent of other work orders. The length of the critical path is therefore three days, and if resources were unlimited, the month's work would be done within that time. So the schedule of work orders is not determined by the critical path, it is determined by the priorities placed on each work order, and by the availability of resources. By prioritizing each work order, and then leveling, a reasonable schedule for the month's work can be obtained.

This illustrates the importance of priority in resource leveling of multiple projects. Most programs provide a priority field for each activity. Some also provide options regarding priority, such as the ability to define whether the highest priority is a high or low number. Other related options involve selection of whether the priority should override other variables when leveling.

MONTHLY MAINTENANCE PROGRAM

CPM-PLAN FOR 100 WORK ORDERS

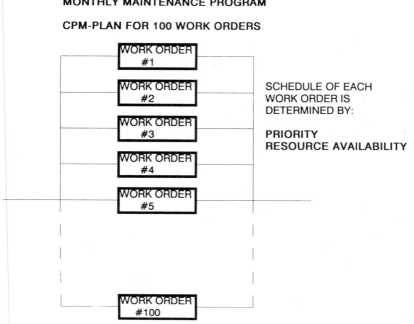

Figure 15.7 A CPM plan for 100 workorders would result in 100 parallel paths thru the network. The critical path is therefore not significant.

Choice of Resource-Leveling Methods

Resource leveling (see Chapters 4 and 7) is a complex calculation in which the program attempts to create a schedule consistent with our ability to provide resources to a project or group of projects. Project management software often provides alternative methods of resource leveling.

Leveling using float only.   This resource-leveling technique attempts to eliminate resource overloads by rescheduling activities between their early and late start dates, i.e., by using total float. The advantage of this technique is, of course, that critical activities are not delayed, nor is project completion. This disadvantage is that the float may not be sufficient to resolve all problems of overloading, so even after leveling, some resource overloads will probably remain.

Nevertheless, this technique is often useful as a first step in the resource scheduling process—to see just how much effect using the float may have.

Full leveling—"resource-limited".   This type of resource leveling calculation will delay tasks as necessary until the resource histogram shows no overloads. Note that the result of this calculation will not be an "optimum schedule," as there may be unused capacity for some resources.

Full leveling—"time-limited".   This type of leveling, usually found in high-end software, preserves the scheduled end-date and attempts to reschedule tasks to minimize the extent of the overload.

Selection of Leveling Procedure

The variables used by a project management program in resource leveling include:

Task priority (high priority gets resources before low)
Amount of float (critical activities get resources before noncritical)
Start date (tomorrow's activities get resources before those starting next
     week)

Some programs allow the user to specify the sequence in which activities should be selected to receive resources. For example, one small project leader may be priority-driven, whereas another is not.

Selection of Resources for Leveling

In some cases, there may be resources that should not be included in the leveling calculation. For example, our manufacturing plant may have only one crane, making the availability equal to one. We want the histogram to show when the requirement for cranes exceeds one, so we prefer to use one as the maximum available. Yet, when leveling, we prefer that the project or activities not be delayed because the requirements for cranes exceeds one. In that case, we can simply rent another crane.

So the ability to specify that resources are to be included and excluded from the leveling calculation is helpful in obtaining a reasonable resource-leveled schedule.

Complex Resource Allocation and Availability

In many programs, resource availability is assumed to be constant over the planning period. This method (referred to as "simple resource availability") is often acceptable for overall planning purposes, but in some cases, additional accuracy is required. Some programs (usually high-end

SIMPLE RESOURCE AVAILABILITY

RESOURCES AVAIL./DAY = CONSTANT

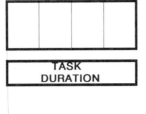

COMPLEX RESOURCE AVAILABILITY

RESOURCES AVAIL./DAY VARIES

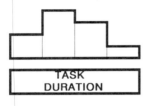

Figure 15.8a   Simple resource availability assumes that a constant number of resources are available over the task's duration. Complex resource availability allows the availability to vary.

or mid-range) permit the availability to be varied over time (see Figure 15.8a). This is referred to as "complex resource availability" and, of course, it provides greater accuracy in the leveling calculation.

In many programs, resource requirements are assumed to be constant over the duration of an activity (see "duration-driven scheduling"). This "simple resource allocation" method is often adequate for overall planning, but, as with resource availability, more accuracy is sometimes required. Again, high end (and some mid-range) programs may provide a "complex resource allocation" capability that permits the requirements to vary over the duration of the activity (see Figure 15.8b).

Export

The capability to export multiple project activity and resource schedule data to a spreadsheet or database may be very useful for analyzing, summarizing, graphing and storing multiple project information.

SIMPLE RESOURCE ALLOCATION

\# RESOURCES REQD./DAY = CONSTANT

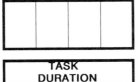

COMPLEX RESOURCE ALLOCATION

\# RESOURCES REQD./DAY VARIES

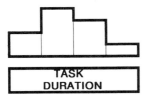

Figure 15.8b   Simple resource allocation assumes that resources are used uniformly over the task's duration. Complex resource allocation allows the utilization to vary.

CUSTOMIZATION AND STANDARDIZATION—
KEY TO MULTIPLE PROJECT EFFICIENCY

As with any management system, success of a multiple-project management method depends on the willingness of each individual to invest the time and effort that the method requires. We can think of this as a challenge to create a method and system that maximizes the return on investment of time of the person who uses it. This return on investment can be maximized by:

Minimizing the time and effort required
Maximizing the benefit received

Customization and standardization of project management software make it possible to create multiple small-project management methods that work, and gain wide acceptance. It is especially important when we consider that, for multiple-project scheduling to be effective, *all projects* (*regardless of size*) *must be included in the schedule.* Therefore the method must pay off even for those who plan very small projects. Fortunately, project management software provides many features that facilitate customization and standardization.

Customization Features

"Customization" refers to the process of configuring the software to meet the specific needs of the particular company and the small-project environment. The idea is to match the specific situation as closely as possible, recognizing that the small project usually must conform to the needs of the organization (not the other way around).

Some useful features found in project management software that facilitate customization are:

Standard Reports

All project management programs provide a variety of standard reports. Often, these standard reports provide options as to how the report is presented and what information is included. Therefore, standard reports can be configured so that they are customized to the small-project leader's situation.

Graphics

Many project management programs provide graphic plotting capability for the network diagram, barchart, progress tracking barchart, etc. As with standard reports, these graphic reports can often be set up with some degree of customization.

Custom Screen Layouts

Any computer user appreciates the ability to work on a screen that has the information that they consider important, and which is arranged in a way they find useful. Many project management programs provide the capability to define screen layouts, such that the appearance, format and contents of the screen are set up specifically to suit a given user. This is especially relevant where multiple small-project management methods must be a good fit with existing practices and preferences.

Custom Reports

The value of any computer program is often measured in a large part by the usefulness of its output. If it produces reports that are easy to read, contain useful information, and are formatted to provide an effective presentation, then the system is likely to be successful. Many project management programs provide the capability to define the format and content of reports, to name the report configuration, and to use it again and again.

Filters

A project model can be thought of as something like a database of project information—on activities, durations, dates, resources, costs, etc. Often, we want to see only part of the information in a database. This is also true of a project model, especially one containing data on multiple projects.

Therefore, many programs provide the capability to define selection criteria that are used to select and display or print only information that meets the selection criteria.

For example, imagine a multiple-project master plan for 100 small projects in a manufacturing plant. We might define a filter that would show just the activities:

In area number 12
Involving the electrical system
Which are on the critical path
Which are scheduled for next week
Which require the resource "electrician"
Which have priority "1"

Like custom reports, filters or selection criteria can often be named, saved, and reused on each new project

Codes

A set of filters often works best when the activities and resources are identified by a customized coding system. Most programs provide the capability to code resource and task information, so that reports can focus on resources, tasks or both.

The Work Breakdown System (WBS) (see Chapter 3) is often used for multiple small projects as a basis for a coding system for activities. For example, the WBS code for an activity might indicate:

Project
Area
System
Type of work
Activity number

Another type of activity code is the Organization Breakdown Structure (OBS). This is often used to indicate the part of the organization that is responsible for the activity.

Likewise, codes can be applied to resources. For example, the resource "June Green" might have a code that identifies June as a toxicologist. Or the resource "welder" may have a code that identifies the welder as being a contract person.

These codes are designed, of course, with an eye toward use with filters to prepare useful and flexible reporting.

Macros

A macro is a method for automating repetitive keystrokes, such as those required to prepare a weekly report. If we must prepare this report every week, and do it for each small project, why not automate that process? A macro, in its simplest form, is nothing more than a file of keystrokes that can be invoked with a single keystroke.

Many programs, project management and otherwise, have a macro capability. Many feature a "learn" or "record" mode in which the macro can be written simply by capturing the keystrokes that we desire to use.

Macros can also be used in a more advanced way, as if they were a programming language. They can be used to control the screen display, provide menus, and even interact with the user with special messages. In fact, they can be used to automate the entire small project planning process, as well as multiple project scheduling, cost estimating, updating, status reporting, and control.

Exporting to Other Software

Spreadsheets have achieved such enormous popularity because of the customization capability they provide. Since project management software can export to spreadsheets, and since spreadsheets can also perform many useful small-project management functions, it makes sense to use spreadsheets in our customization process.

Project management software can often be customized to set up a special, user-defined report that is then exported to a spreadsheet or database for further processing. For example, a cost estimate for direct material and labor can be set up in a special format in the project management software, then exported to a spreadsheet in which the final touches are added and the presentation output prepared.

Spreadsheet operations can also be automated to a high degree with macros. We can easily set up a multiple small-project management system that uses macros to automate both the project management and spread-sheet operations.

Standardization Features

"Standardization" refers to the process of establishing standard project models (or parts of project models), to be used on each and every small project. The idea is to minimize the time and effort required to plan, schedule and estimate each small project, as well as to prepare necessary periodic reports.

With standardized project models, the work required for planning a new small project is reduced to this:

Retrieve the appropriate master file
Edit tasks, resources as required
Prepare reports

And all of these tasks can be automated with macros. The idea is that we set up one or many standard project models, template plans, if you will, so that the work required to plan a new small project is reduced to editing a template plan. The specific elements that can be used to achieve standardization include:

Standard Resource and Cost List

Since multiple small projects tend to draw from the same pool of resources, it makes sense to prepare a standard list (or lists) of resources. These resources can be individual resources (e.g., Jim Jones) or resource categories (e.g., electrical engineer). Each new project that is planned then draws resources from this list, rather than redefining the list each time. If necessary, resources can be added to the list to fit the needs of a specific project, likewise, any one project may not use all the resources on the list. The resource list may also include equipment (e.g., CAD workstation) and material costs (e.g., electrical cable). Use of this standard list will also assure consistency with custom reports, codes and filters.

Standard Calendars

Most if not all of the multiple small projects will use the same workdays and holidays. There will probably be a standard calendar (or calendars) that all projects can use.

In addition, many programs provide the capability to define a calendar for each specific resource. So the standard resource list and calendar go together, and should be used together.

Standard Milestones

Milestones have proven to be a useful planning tool for communication of schedule status. When reporting the status of multiple small projects, it is often helpful to have milestone reports for use by management. Why not, then, have standard milestones that are automatically part of each small project plan? These milestones, along with the standard resources and calendars, can be loaded as part of the template plan used to build each small project.

Standard Plans

When we look at the type of work done in the multiple small-project environment, we often find that many projects are essentially repetitive. There are just not that many different ways to conduct a lab test, design an electrical system, program a computer, or maintain a pump. It is, therefore, often a practical approach to prepare a group of standard project plans (template plans) whose purpose is to provide a starting point for planning similar projects.

For example, if we are planning various lab testing operations, we might start the plan for each one by retrieving a template plan that is close to the project at hand. We then edit that plan by adding, deleting, and changing activity and resource data, until we have what we want. Standard macros, filters and reports are available to speed up the process.

Standard Reports

The management of multiple small projects is made much easier when the plans, schedules, cost estimates, resource reports, cost reports, and progress reports all look the same. Consistency and uniformity are the goals of all managers in this environment.

Therefore, the ability of project management software to define custom reports and then provide these reports consistently on all projects, is most important.

Macros—Preparing Customized Application Programs

Let's take a more detailed look at how a macro language can be used to automate the planning, scheduling and control of multiple small projects.

1. The project management program is loaded using a command that invokes a macro. As the program loads into memory, a customized screen appears with the message: "Welcome to the XYZ Company Project Planning Program. What type of project do you wish to plan?"

2. The user then selects a type of project from a menu. A template plan is retrieved, containing standard activities, milestones, resources, calendars, filters, and codes.

3. The macro (still running) guides the user through the process of inputting the project name, start date, leader, etc.

4. Macros are also used to invoke resource leveling and other operations necessary to preparing the plan and schedule.

5. A reporting macro is invoked, using a "Function key." A screen appears, asking: "You are about to prepare a report. What type of report do you want?"

The user may then choose from a menu of reports including: Cost, Schedule, Planning, Milestone, Resource Requirements, Progress etc.

The report macro then asks if the user wants to:

View the report
Print the report
Plot a graphics report
Export the report to a spreadsheet

A similar macro-driven procedure can be used for updating and status reporting, as well as multiple project scheduling.

6. To prepare a management report, the multiple project manager loads the spreadsheet program. He or she invokes a macro that retrieves the exported project management file, and processes that data into various spreadsheets that are used to create the required charts, graphs and reports. The macro may prompt the user for additional input where appropriate, as well as provide help and explanation screens and boxes.

COMBINING PROJECT MANAGEMENT WITH GENERAL PURPOSE PC SOFTWARE

In spite of the flexibility offered by project management software, there are many situations in which general-purpose software is required. By general purpose software, we mean software not designed specifically for project management such as:

Spreadsheets (e.g., Lotus 123, Quattro Pro, Excel)
Databases, including "flat-file" types (e.g., dBase, Foxbase, Q&A)
Management graphics (e.g., Harvard Graphics, Freelance)
Word processors (e.g., WordPerfect, Microsoft Word)

The combination of project management software with general purpose software creates a capability to produce almost any kind of analysis or report that could be required for the management of multiple small projects. Let's look at some examples of situations in which these types of programs can be used together effectively.

Some project management programs are written with the specific intention of being compatible with general purpose software. An example of this is the high-end program Open Plan, which is compatible with dBase (and Foxbase) software. This permits the development of sophisticated, customized applications in which project data (such as costs), which are best handled in a database, can be truly integrated with other data (such as dates), which are best handled in a CPM-type program.

Cost Estimating

In a cost estimating application, it is often desirable to use a project management program to estimate direct costs such as labor, equipment and materials using project management software. However, it is often required that a cost estimate be presented to management in a certain format. Furthermore, it may be necessary to complete a cost estimate by including certain calculations that are not possible in project management software.

These requirements can be easily met using the export capability that is readily available in most low-end project management programs. Using the "customized and standardized" project model, plans are prepared for each small project. Macro routines are used to create and print customized and standardized reports, showing the material and labor costs required to do the work. Another type of report that the project management program can create is one showing the cashflow requirements over the project's life. These cost estimate reports can be automatically prepared in the form of a spreadsheet file (e.g., a .WK1 file).

Now imagine a spreadsheet program into which the cost estimate is imported. The spreadsheet, also macro-driven, collects the cost data it needs, then automates the process by which additional cost elements are calculated. These might include:

Currency exchange calculations
Calculating escalation
Calculating general overhead costs
Including company-specific charges, such as services from other departments
Contingency calculations

Of course, the estimating spreadsheet would assure that charts and graphs describing the estimate are also prepared.

Earned-Value Progress Measurement

Another situation in which exporting to a spreadsheet is useful is in the preparation of earned-value calculations. Although many project management programs, even at the low-end, have some earned value capabilities, these may not provide all the desired functions. For example, it might be desirable to capture actual costs and workhours from another inhouse program, and include these in the calculation of productivity. This often can be done more easily in a spreadsheet than in the project management program.

This type of application might take updated activity data, showing progress and other status-related data, and (again using a macro) prepare a report for export to the spreadsheet. In the spreadsheet, we rearrange this captured data as necessary, and merge it with the actual cost and hours (input by hand or transferred from another program). The result is a productivity analysis, and various management reports that could easily include tracking curves.

Status Reporting

This is a similar type of operation to the one described above, but broader in scope. For example, a service firm might want to analyze progress, costs, expenditures, and commitments in reference to the contract price. The goal here is to analyze the profitability status of the project.

In another case, specific requirements for status report format and content might dictate the use of exporting to a spreadsheet. And, of course the rich assortment of charts and graphs that most spreadsheets provide permit all kinds of tracking curves, pie charts, histograms, 3-D plots and so on—all of which are most useful for management status reporting.

PERT-Based Probabilistic Analysis of Cost and Schedule

The PERT calculation described in Chapters 3 and 6 is also an application of export functions. One can prepare a project plan in the project management software, and then export a report showing the most likely duration and dates for each activity. The spreadsheet can then be used to prepare a report in which the maximum and minimum values are input or calculated, thereby permitting the calculation of the "50/50" value of duration of each activity as well as of the entire project.

Likewise, cost data for each resource category can be prepared in the project management program, then exported to a spreadsheet for a PERT cost analysis. (Note that a few programs, such as Superproject Expert and Open Plan, do provide a PERT function within the project management operation.)

The spreadsheet can also be used to expand the basic PERT calculation into a more complete and sophisticated analysis of the probability functions of cost and schedule. Such an analysis could predict the risk of overrun and schedule delay, as well as quantify accuracy and contingency.

CHAPTER SUMMARY

In this chapter, we have explored the many ways in which personal computers can be used to perform the calculations and analyses of multiple small-project management. Indeed, even low-end, inexpensive hardware and software make it possible to implement all of methods described in this book. If there is a most frequent mistake made in viewing PC software for project management, it is underestimating what it can do.

This chapter described the specific features to look for in PC software to help manage multiple projects. With the right software, set up the right way, and the right practices and procedures, the management of multiple small projects can be effective and successful.

Index